```
QH                        100776
105
.T4
E57    Rabalais, Nancy N.,
       Enviromental studies of a marine
       ecosystem
```

ENVIRONMENTAL STUDIES OF A MARINE ECOSYSTEM
South Texas Outer Continental Shelf

ENVIRONMENTAL STUDIES OF A MARINE ECOSYSTEM
South Texas Outer Continental Shelf

Edited by R. Warren Flint and Nancy N. Rabalais

UNIVERSITY OF TEXAS PRESS, AUSTIN

Publication of this book was assisted by the Hooks Contingency Fund.

Copyright © 1981 by the University of Texas Press
All rights reserved
Printed in the United States of America

First Edition, 1981

Requests for permission to reproduce material from this work should be sent to
Permissions
University of Texas Press
Box 7819, Austin, Texas 78712

LIBRARY OF CONGRESS CATALOGING IN PUBLICATION DATA

Main entry under title:
Environmental studies of a marine ecosystem.
  Bibliography: p.
  Includes index.
  1. Marine ecology—Texas. 2. Marine ecology—Mexico, Gulf of. 3. Continental shelf—Texas. 4. Continental shelf—Mexico, Gulf of. I. Flint, R. Warren, 1946– II. Rabalais, Nancy N., 1950–
QH105.T4E57   574.5'2636'0916364   80-28275
ISBN 0-292-72030-0

CONTENTS

Contributors xiii
Foreword xv
Preface xvii
Overview xix

1. Introduction 3
2. Marine Pelagic Environment 15
3. Pelagic Biota 36
4. Marine Benthic Environment 68
5. Benthic Biota 83
6. Ecosystem Characteristics 137

APPENDICES

A. Overall Base Line Results 157
B. Maps of Variables' Geographic Distributions 199

References 221
Index 231

FIGURES

1. Conceptual model of continental shelf ecosystem   5
2. Bathymetric map of the Gulf of Mexico   7
3. Map of south Texas continental shelf and sampling sites   8
4. Time-depth plots of water temperature and salinity   20
5. Plot of bottom water temperature and salinity variability   21
6. Water temperature cross-section along Transect II   22
7. Relationship between salinity and secchi depth   24
8. Contours of chlorophyll *a*, salinity, and temperature   25
9. Relationships between chlorophyll *a*, salinity, and freshwater sources   26
10. Nutrients in surface waters   30
11. Phosphate concentrations with depth   31
12. Depth profile of methane, ATP, and chlorophyll *a*   34
13. Temporal patterns of inner-shelf chlorophyll *a* components   37
14. Temporal patterns of mid-shelf chlorophyll *a* components   39
15. Temporal patterns of outer-shelf chlorophyll *a* components   40
16. Temporal patterns of carbon-14 uptake   41
17. Primary production for inner-shelf waters   42
18. Comparison of seasonal abundances of classes of phytoplankton   44
19. Comparison of seasonal abundances of species or genera of phytoplankton   45
20. Depth profiles of light, transmissometry, chlorophyll *a*, and ammonia nitrogen   47

[viii] Figures

21. Spatial variation in tar concentration 54
22. Temporal variation of zooplankton abundance variables 56
23. Temporal variation of zooplankton community structure variables 57
24. Frequency of occurrence of female copepod species 59
25. Zooplankton hydrocarbon concentrations 63
26. Station groups of similar sediment characteristics 73
27. Percentage of sediment n-alkanes 77
28. Methane variations in surficial sediment 78
29. Interstitial hydrocarbon concentrations 79
30. Interstitial ethane concentrations 80
31. Interstitial propane concentrations 81
32. Effect of crude oil on fungal growth 86
33. Relation between sediment aerobic heterotrophic bacteria and depth 89
34. Relation between oil-degrading sediment bacteria and alkanes 90
35. Distribution of permanent meiofauna 92
36. Distribution of nematodes 93
37. Results of benthic infauna community ordination 98
38. Plots of infaunal community characteristics 99
39. Infaunal species frequency of occurrence 100
40. Discriminant analysis of environmental variables associated with infauna 102
41. Benthic environmental variables 103
42. Epifaunal communities by cluster analysis 108
43. Species dendrogram of epifauna 109
44. Community structure variables for epifauna in winter 110
45. Community structure variables for epifauna in spring 111
46. Community structure variables for epifauna in fall 112
47. Demersal fish communities by cluster analysis 115
48. Discriminant analysis of environmental variables associated with demersal fishes 120
49. Discriminant analysis of fish abundance 121
50. Distribution of n-alkanes in brown shrimp 126
51. Comparison of hydrocarbons in brown shrimp 128
52. Relationship of pelagic and benthic densities to water depth 138
53. Temporal profiles for ammonia nitrogen concentrations by depth 141

54. Correlational model of ecosystem variables   144
55. NOAA statistical reporting area   148
56. Conceptual trophic model for brown shrimp fishery   149
57. Plot of brown shrimp fishery yield   152

APPENDIX A

A-1. Sampling sites   162
A-2. Comparison of empirical and normal curve confidence intervals   167
A-3. Cells demonstrating unbalanced data collections   169

APPENDIX B

B-1. Surface water silicate   200
B-2. Surface net chlorophyll   201
B-3. Surface net phaeophytin   202
B-4. Bottom water phosphate   203
B-5. Bottom water dissolved oxygen   204
B-6. Bottom water total chlorophyll   205
B-7. Bottom water propene   206
B-8. Bottom water ethene   207
B-9. Copepod total density   208
B-10. Sediment mean grain size   209
B-11. Sediment grain size standard deviation   210
B-12. Sediment total organic carbon   211
B-13. Sediment Delta ^{13}C   212
B-14. Total sediment bacteria   213
B-15. Total meiofauna species   214
B-16. Total meiofauna density   215
B-17. Nematode density   216
B-18. Harpacticoid density   217
B-19. Infauna species   218
B-20. Infauna density   219

TABLES

1. Station locations and depths   9
2. Participants by work element and institution   11
3. Study areas focused upon in data integration   12
4. Low molecular-weight hydrocarbon concentrations   32
5. Average total hydrocarbons in seawater by station   35
6. Mean abundances of selected neuston taxa   50
7. Concentrations of trace elements in zooplankton   64
8. Seasonal concentrations of trace elements in zooplankton   66
9. Mean and standard deviation for several sediment variables   74
10. Summary of sediment Delta ^{13}C and organic carbon   75
11. Fungal abundance in surficial sediments   85
12. Growth of fungal isolates in crude oil   85
13. Summary of benthic bacterial populations during 1977   88
14. Abundance of major taxa of the meiobenthos   95
15. Composition of major faunal groups   101
16. Total abundance and occurrences of demersal fishes   117
17. Seasonal occurrence of common fishes   118
18. Spatial distribution of common fishes   119
19. Heavy hydrocarbons in macroepifauna and macronekton   124
20. Ranges of hydrocarbons in macronekton   130
21. Selected hydrocarbon variables in macronekton   130
22. Concentrations of trace metals in demersal fish   132

Tables

23. Concentrations of trace metals in penaeid shrimp   134
24. Comparison of Atlantic Ocean and Gulf macrobenthos   152

APPENDIX A

A-1. Hypothetical results illustrating Table A-5 format   158
A-2. Station groups by depth for 12 stations   163
A-3. Station groups by depth for 25 stations   164
A-4. Two-factor analyses strategy   171
A-5. Distributional characteristics for ecosystem variables of the south Texas shelf   178

CONTRIBUTORS

Alexander, Steve K., Moody College of Marine Sciences, Texas A&M University, Galveston, Texas 77550

Behrens, E. William, Geophysical Laboratory, The University of Texas Marine Science Institute, Galveston, Texas 77550

Bernard, Bernie B., Department of Oceanography, Texas A&M University, College Station, Texas 77843 (Present Affiliation: School of Geology and Geophysics, University of Oklahoma, Norman, Oklahoma 73019)

Boothe, Paul N., Department of Oceanography, Texas A&M University, College Station, Texas 77843

Brooks, James M., Department of Oceanography, Texas A&M University, College Station, Texas 77843

Flint, R. Warren, Port Aransas Marine Laboratory, The University of Texas Marine Science Institute, Port Aransas, Texas 78373

Giam, Choo-Seng, Department of Chemistry, Texas A&M University, College Station, Texas 77843

Godbout, Robert C., Port Aransas Marine Laboratory, The University of Texas Marine Science Institute, Port Aransas, Texas 78373

Holland, J. Selmon, Port Aransas Marine Laboratory, The University of Texas Marine Science Institute, Port Aransas, Texas 78373 (Present Affiliation: Division of Fisheries Rehabilitation, Enhancement, and Development, Juneau, Alaska 99803)

Kamykowski, Daniel L., Port Aransas Marine Laboratory, The University of Texas Marine Science Institute, Port Aransas, Texas 78373 (Present Affiliation: Department of Marine Science and Engineering, North Carolina State University, Raleigh, North Carolina 27650)

McEachran, John D., Department of Wildlife and Fisheries Sciences, Texas A&M University, College Station, Texas 77843

Neff, Grace, Department of Chemistry, Texas A&M University, College Station, Texas 77843

Park, E. Taisoo, Moody College of Marine Sciences, Texas A&M University, Galveston, Texas 77550

Parker, Patrick L., Port Aransas Marine Laboratory, The University of Texas Marine Science Institute, Port Aransas, Texas 78373

Pequegnat, Linda H., Department of Oceanography, Texas A&M University, College Station, Texas 77843

Pequegnat, Willis E., Department of Oceanography, Texas A&M University, College Station, Texas 77843

Powell, Paul, Department of Microbiology, The University of Texas at Austin, Austin, Texas 78712

Presley, Bobby Joe, Department of Oceanography, Texas A&M University, College Station, Texas 77843

Rabalais, Nancy N., Port Aransas Marine Laboratory, The University of Texas Marine Science Institute, Port Aransas, Texas 78373

Scalan, Richard S., Port Aransas Marine Laboratory, The University of Texas Marine Science Institute, Port Aransas, Texas 78373

Schwarz, John R., Moody College of Marine Sciences, Texas A&M University, Galveston, Texas 77550

Smith, Ned P., Port Aransas Marine Laboratory, The University of Texas Marine Science Institute, Port Aransas, Texas 78373 (Present Affiliation: Harbor Branch Foundation, Fort Pierce, Florida 33450)

Szaniszlo, Paul J., Department of Microbiology, The University of Texas at Austin, Austin, Texas 78712

Turk, Phil, Moody College of Marine Sciences, Texas A&M University, Galveston, Texas 77550

Venn, Cynthia, Department of Oceanography, Texas A&M University, College Station, Texas 77843

Winters, J. Kenneth, Port Aransas Marine Laboratory, The University of Texas Marine Science Institute, Port Aransas, Texas 78373

Wohlschlag, Donald E., Port Aransas Marine Laboratory, The University of Texas Marine Science Institute, Port Aransas, Texas 78373

Wormuth, John H., Department of Oceanography, Texas A&M University, College Station, Texas 77843

Yoshiyama, Ronald, Port Aransas Marine Laboratory, The University of Texas Marine Science Institute, Port Aransas, Texas 78373 (Present Affiliation: Environmental Sciences Division, Oak Ridge National Laboratory, Oak Ridge, Tennessee 38730)

FOREWORD

This study of the south Texas outer continental shelf was conducted on behalf of the Bureau of Land Management, U.S. Department of the Interior, and with the close cooperation of personnel of that agency. The overall program included information on (1) geology and geophysics by the U.S. Geological Survey; (2) fisheries resources and ichthyoplankton populations by the National Marine Fisheries Service, National Oceanic and Atmospheric Administration, U.S. Department of Commerce; and (3) biological and chemical characteristics of selected topographic features in the northern Gulf of Mexico by Texas A&M University. The data resulting from this investigation represent the environmental background existing before major petroleum exploration and development commence in the area. The central goal of these and other environmental quality surveys of continental shelf areas is the characterization and protection of the living marine resources.

This investigation was the result of the combined efforts of scientists and support personnel from several universities. The hard work and cooperation of all participants are acknowledged.

PATRICK L. PARKER
The University of Texas Marine Science Institute
Port Aransas, Texas

PREFACE

The present concern about the rate of fossil fuel consumption and dependency upon imported oil to supply current U.S. demands has resulted in a greater focus of interest by both the government and oil companies on the U.S. continental shelf for increased domestic production. The 1969 National Environmental Policy Act identifies the U.S. Department of the Interior as the agency responsible for protecting the marine environment of the continental shelf during periods of exploration and exploitation of natural resources. To obtain information upon which to base decisions concerning the orderly development of these resources while also striving to protect the environment, the Bureau of Land Management (BLM), an agency of the Department of the Interior, established a marine environmental studies program for the outer continental shelf.

This book presents the results of three years of field studies and data collection on the south Texas outer continental shelf in one of the BLM programs, integrating the information obtained into a statement of the ecosystem characteristics of this shelf area. The intent of the contributors is to provide initial information environmental managers need to make sound decisions concerning natural resource exploitation in continental shelf waters. Besides a general ecosystem description, this book presents those environmental relationships that exist and those specific environmental characteristics (variables) that are most important for predicting, assessing, and managing impacts on the south Texas outer continental shelf ecosystem.

On 3 June 1979, while this book was in preparation, a well blowout occurred at the IXTOC I drilling site in the Bay of Campeche off the Mexican coast in the southwestern Gulf of Mexico. The events that followed this major disturbance to the marine environment of the Gulf, as massive oil slicks entered U.S. waters, emphasized the value of this study program in establishing base line conditions and ecosystem characteristics. Federal agencies associated with the national oil spill response team that monitored the IXTOC I spill and developed a damage assessment research plan were able to use data from this study to identify critical components and important variables of the shelf environment that could be used to detect ecosystem change from the spill impact. The editors only hope that the reasons for conducting the south Texas outer continental shelf research program will not be forgotten. Now that the opportunity exists to evaluate the actual impact of a major perturbation of natural resource exploitation, decision makers need to take full advantage of the extensive data base available to fill numerous information gaps so that future decisions involving any shelf environment and resource exploitation can be made without a feeling of apprehension and uncertainty.

Special acknowledgment is accorded the scientists who served as program managers for the duration of the research program detailed in the following pages, including Robert S. Jones, Robert D. Groover, and Connie R. Arnold. Acknowledgment is also given to Richard Casey, Jerry Neff, William Haensly, Patricia Johansen, Chase Van Baalen, Samuel Ramirez, Helen Oujesky, William Van Auken, and Neal Guntzel for their overall scientific contributions to the south Texas outer continental shelf program, even though they did not participate in the final data synthesis. Thanks are also extended to D. Kalke and L. Tinnin for preparation of the final manuscript. We further thank the Bureau of Land Management for providing financial support.

R. WARREN FLINT

OVERVIEW

The broad continental shelf of south Texas supports valuable commercial and sport fisheries, particularly of penaeid shrimp, along with potential sites for exploration and exploitation of oil and gas resources. An intensive, multidisciplinary three-year study (1975–1977) to characterize the temporal and spatial variation of both living and nonliving resources of the area was designed to provide initial information needed by environmental managers to make sound decisions concerning natural resource exploitation. The synthesis and integration of the data gathered resulted in an encompassing description of the physical, chemical, and biological components of the system, identification of the temporal and spatial trends that best represent the ecosystem, along with mathematical descriptions of unique relationships that would serve as "fingerprints" by which subsequent changes or impacts could be measured, particularly those related to oil and gas development activities. The study included the pelagic environment, its physical characterization and biotic composition and productivity; the benthic habitat, its physical setting and biotic composition; and inherent natural petroleum hydrocarbon and trace metal levels in selected portions of the physical and biological components, both pelagic and benthic.

As research priorities were reassessed and additional information determined necessary, the study was amended appropriately to meet evolving study objectives. Sampling schemes varied from year to year and according to particular components of the overall study. In general, the study area was traversed by four transects perpendicular to shore, each with six or seven stations distributed from 10 to 130 m offshore. The number of stations sampled varied with the year, the study element, and the collection period. Samples were taken seasonally along all four transects during all three years. Monthly sampling was conducted on Transect II during 1976. Additional sites included stations near hard-bottom features, Hospital Rock and Southern Bank, and a rig monitoring station. A summary of the highlights of this study follows.

The Texas shelf environment is a complex interaction of adjacent land masses, coastal waters near shore influenced by estuarine systems and their inherent high productivity, riverine input (in particular from the Mississippi River), and dynamics of open Gulf waters. The climate of south Texas is subtropical and semiarid with an average yearly rainfall of 70 cm. Because of these conditions, not only along the coast but landward for more than 100 miles, no major streams flow to the Gulf of Mexico along the coast between Aransas Pass Inlet and the Rio Grande, 125 miles to the south. The general circulation of air near the Gulf surface over the south Texas coastal region follows the sweep of the western extension of the Bermuda high-pressure system throughout the year. Relatively high surface water temperatures of the Gulf bring about a great warming and an increase in moisture content in overlying air masses. Water mass distribution in open Gulf waters results from inflow through the Yucatán Channel, outflow through the Straits of Florida, surface conditions created by local air-sea exchange processes, and internal mixing of three well-defined water masses—Gulf basin water, a layer of the Antarctic intermediate water, and a mid-Atlantic element. The hydrography is a mixture of these elements and is important as the basic setting for the resultant biological communities, which are a reflection of it.

In surface water layers from 10 to 100 m deep across the south Texas shelf there is a strong cross-shelf temperature gradient during midwinter that disappears with seasonal heating until the surface water is spatially isothermal at 29°C by late summer. The winter gradient produces the lowest values (to 14°C) over the inner shelf and minimum values (19°C to 20°C) over the outer shelf. Vertical stratification, nearly absent in shelf waters during the winter, is well developed in the summer, being more prevalent with depth. Shelf salinities are high most of the year, except in a short period in spring and early summer when a plume of Mississippi River water may cover the entire shelf, lowering salinities through the uppermost 20- to 30-m water depth. There is suggestion of occasional upwelling of deep Gulf water onto the shelf. An aspect of prime importance, particularly to the pelagic biological communities and the benthos, is the extreme variability of shallow waters and the contrasting stability of deep waters in both temperature and salinity. In shallow waters, salinity is almost totally influenced by local rainfall and riverine input. Annually a plume of Mississippi River water moves westward and southwestward along the northern rim of the Gulf of Mexico during spring. This plume is especially pronounced along the coast but at times covers the entire shelf. The shelf is thus divided into three zones: (1) an inshore zone dominated by Texas riverine input; (2) a middle zone in which Texas freshwater sources and the Mississippi River are influential, with a gradation from one to the other with increased distance offshore; and (3) an offshore zone dominated by Mississippi River discharge.

The longshore currents toward the southwest dominate October through March and are responsible for advective transport of Mississippi River waters along the northwestern rim of the Gulf of Mexico at a time when discharge is the greatest. Between June and September the longshore component weakens and reverses over short time scales to periodically produce perpendicular movements of water across the shelf. Surface currents near shore are influenced by locally prevailing winds. These water movements influence the transport of nutrients, heat, suspended solids, and planktonic life.

Study of the pelagic biota shows that Texas shelf waters are extremely high in annual phytoplankton productivity. Primary production in inner-shelf waters is bimodal annually with peaks in spring and fall. There is a cross-shelf gradient of chlorophyll a concentrations with a peak inshore and a steep drop offshore. Although not as strong, a north-south gradient for chlorophyll a is also on the shelf. The northern part of the shelf is higher in chlorophyll a at the surface and at half the depth of the photic zone than is the southern part. There is no north-south gradient of chlorophyll a in the bottom waters, indicating a lack of mixing on the outer shelf. The highest concentrations of chlorophyll a are often in the bottom waters, especially at shallow stations characterized by a pervasive nepheloid layer. In this layer, peak chlorophyll levels (primary producer biomass), adequate light transmittance, and evidence of nutrient regeneration lead to occurrence of photosynthesis in bottom waters.

The phytoplankton community is complex but relatively consistent and a reflection of different water masses on the shelf over annual cycles, with temporal changes in community structure related to light intensity, day length, temperature, salinity, stratification, wind, and nutrient sources. Geographical trends in the phytoplankton are usually related to water depth and distance from shore, with the highest abundances along the inner shelf. High phytoplankton numbers in spring are correlated with riverine input and nutrient maxima.

Neuston, the biota living on or just beneath the surface film of marine waters, varies considerably in abundance, either as total numbers or dry weight, as well as in taxonomic composition. Part of the variability is a result of diel vertical migration; the remainder is a reflection of environmental heterogeneity. Cross-shelf variation in the distribution of some taxa, particularly the larval decapod crustaceans, occurs annually and is related to benthic distribution patterns of adults and to estuarine influences. Neuston is significantly correlated with the density of microtarballs. This relationship may be accounted for by windrowing effects of surface water circulation or by the potential food source of epibiotic species on well-weathered petroleum products.

Zooplankton biomass and total density decrease with distance offshore. A few species, primarily female copepods, dominate the zooplankton density. There is considerable north-south variability, suggesting the occurrence of pulsing input to the system that encourages zooplankton production but is so limited that the entire length of the south Texas shelf is not uniformly affected. The patchy distribution of zooplankton may be related to low salinity input from bay systems. Evidence for estuarine influence is the increased numbers of *Acartia tonsa*, a calanoid copepod abundant in bays and estuaries of the Gulf of Mexico, in the spring when salinity is low at stations near shore and mid depth on the northern half of the shelf. Salinity is related to several zooplankton variables in shallow waters but most frequently correlates with zooplankton variation at mid-depth points on the shelf. The implied relationships between zooplankton variables and salinity on the middle of the shelf may indirectly reflect a response of the zooplankton to changes in primary production, which has been shown to be commonly associated with salinity changes in neritic waters. The direct relationship of zooplankton to phytoplankton in deeper waters reflects a close dependence of zooplankton on phytoplankton. The offshore zooplankton population may be controlled by food availability, while zooplankton populations near shore may be controlled by predation.

The general feature of the sea bottom is a broad ramp-like indentation on the outer shelf between two ancestral deltaic bulges, the Colorado-Brazos in the north, seaward from Matagorda Bay, and the Rio Grande in the south. The sea floor is characterized by sand-sized sediments on the inner shelf that decrease in abundance seaward. Sand is transported seaward from the high-energy zone of the innermost shelf. The encroachment of sand particles onto the Texas shelf from the north suggests a regional southward movement of sediment.

Within the study area's deepest waters (106 to 134 m), sediments are characterized by silty (30%) clay of a very uniform texture, occasionally coarsened by winnowing during the early spring. A slightly coarser, more variable silty clay is associated with northern areas of the south Texas shelf in waters between 65 and 100 m deep. These are transition areas between the clayey sediments found deeper and the silty sediments found between depths of 36 and 49 m in the northern part of the shelf. Farther landward (at 18- to 22-m depths in the northern half and 25- to 37-m depths in the southern half) are the most variable inner-shelf sandy muds. A similar area with greater variability, at least partly because of coarse sand with some gravel, is located at depths between 47 and 91 m on the Rio Grande delta. On the inner shelf, there is a suggestion that seasonal coarsening occurred in 1977, perhaps related to hurricane-generated waves between spring and fall.

One of the major focuses of this multidisciplinary study was characterization of the subtidal benthic habitat. Unlike the water masses and associated biota that are in continual motion, the benthos is relatively stationary and thus serves as a barometer measuring changes that occur in localized areas. Natural variation in the benthos or the transfer of materials through the community or both are important in understanding the essential links in the trophic dynamics of the Gulf of Mexico.

The benthic community was studied here according to components determined categorically by taxa, size fractions, or relative position in the benthos. These components include microorganisms, both fungal and bacterial; organisms living in the sediments, both the meiofauna (< 0.5 mm) and the macrofauna (> 0.5 mm); and organisms living above the sediments but closely associated with it, the invertebrate epifauna and the demersal fishes.

Marine fungi are present in benthic sediments in low numbers in the late winter but significantly increase through the year to fall. The abundance of fungi appears to be controlled by the replenishment of inoculum seasonally from the water column, which in turn depends on deposition in the water column from continental air masses and the local availability of organic carbon. Fungi are short lived in sediments where available carbon is limited. A pattern of increasing numbers is paralleled by an increase in numbers of taxa. Over 50% of the benthic fungi are capable of assimilating crude oil to overcome carbon limitations, but this oil degradation potential decreases offshore. It is reasonable to presume at least some fungal oxidation of intrusive petroleum contamination would occur in the area.

Marine aerobic heterotrophic bacteria are found in sediments in numbers from 4.6×10^4 to 1.3×10^6/ml wet sediment. Numbers are highest during spring and lowest during winter. Greatest populations occur in shallow waters and decrease as depth increases offshore. Benthic bacteria appear to respond to the high input of organic carbon into the sediments during periods of peak productivity in the overlying water column in spring and when sediment temperatures are lowest in winter. Hydrocarbon-degrading bacteria are present in sediments throughout the area and are more numerous near shore with decreasing numbers offshore. They are significantly correlated with the total alkanes in the sediments. Benthic bacteria are capable of degrading all n-alkanes (C_{14} to C_{32}) but exhibit a preference for lower hydrocarbons (C_{14} to C_{20}). Stimulation of total anaerobic heterotrophic bacteria and hydrocarbon-degrading bacteria by the addition of crude oil to the sediment occurs over most of the shelf.

The meiofauna are those organisms smaller than 0.5 mm but larger than 0.1 mm. This somewhat arbitrary definition of size is used to distinguish these small metazoans from the larger macrofauna of the benthos. Further delineation excluding the young of the macrofauna and including only species that at the adult stage fit into the size and taxonomic categories (i.e., the permanent meiobenthos) provides a better operational definition in terms of sampling methods. It also provides meiofauna a natural grouping having certain biological characteristics, differing from the macrofauna in their reproductive capacity and general metabolism as well as in the ecologic niche they fill. Meiofaunal populations diminish with increasing depth on the Texas shelf. Consistently, southern inshore shelf areas support the highest populations and northern areas the lowest. Populations of the deepest waters in the middle of the south Texas shelf are almost as great as those of the shallowest points in that area. In contrast, populations at the deepest stations on other parts of the shelf are only a small percentage of those of the shallowest stations. Nematodes are the most abundant meiofaunal taxa, averaging 93% of the total abundance of the permanent meiofauna. There is a marked increase in nematodes when the sand content of the sediment is 60% or more by weight.

The macroinvertebrate infauna data cluster into geographical areas similar to sediment distributions, and community variables exhibit trends consistent with these clusters. The number of species is highest in shallow waters but significantly drops mid depth on the shelf.

Density is also greatest for the shallow areas and decreases in deeper waters on the shelf. These variables result in high species diversity measures for the inner shelf. The highest diversity, however, is seen mid depth on the southern extremes of the south Texas shelf. The shallow waters are characterized by a few dominant fauna in contrast to the more evenly distributed populations offshore. Specific faunal assemblages describe the geographical areas delineated. The species groups represent shallow, mid-shelf, and deep water fauna as well as ubiquitous fauna. Polychaetes are by far the dominant taxa throughout the shelf.

Analysis of physical variables associated with the benthos geographical areas indicates that there are environmental differences between them. Water depth is the dominant variable accounting for benthic community groupings on the shelf. Additionally, the sediment properties of the sand-to-mud ratio, the sediment grain size deviation, and the percentage of silt account for variation between faunal provinces. Factors related to water depth, benthos food availability, and bottom water variability along the depth gradient must also be considered. Chlorophyll a concentrations are highest and also most variable in shallow waters, where the highest densities of infauna occur. Lower concentrations of primary producers with less variable abundances in deeper waters are associated with lower densities of infauna and more evenly distributed population numbers within these assemblages. Temperature and salinity are also most variable at shallow depths, but variability decreases with increasing water depth. The shallow benthic habitat is more variable and less predictable in terms of environmental change than the deep benthic habitat and thus conducive to dominance by a few fauna.

As with the macroinvertebrate infauna distributions, depth is the most apparent factor controlling epifaunal distributions. The shelf is divided into two major regions based on benthic epifaunal species assemblage patterns: (1) a shallow zone (10 to 45 m) with variable bottom water temperature (10°C to 29°C) and salinity (30‰ to 37‰) and sandy sediments and (2) a deeper region (> 45 m) with a more stable temperature (15°C to 25°C) and salinity (35‰ to 37‰) and high clay content. Many of the species characteristic of the shallow shelf are motile decapod crustaceans found in inlets, bays, and coastal waters in summer and early fall. Numerous species, each in low abundance, characterize the outer shelf assemblage.

The demersal fish populations also align with shelf depth into three distinct faunal provinces, with seasonal migration patterns influencing the species' associations. The shallow shelf zone exhibits little species diversity throughout the year, but there are especially high numbers of each species in winter and spring. The faunal association near shore dissipates during late summer or autumn when shallow shelf water temperatures are highest. Mid-depth associations are the most diverse and most stable throughout the year. There is considerable species "shuffling" during the year in all faunal zones, suggesting that species-dominated communities do not persist. Analysis of physical variables associated with demersal fish populations indicate that mean sediment grain size, salinity, silt percentage, and other sediment characteristics account for the environmental differences between the depth-related stations.

The minimal presence of hydrocarbons in Texas shelf waters and sediments indicates that the area is relatively pristine, and those hydrocarbons observed can be attributed primarily to natural sources. Natural sources include both primary production and bacterial production in highly active water layers near the air-water interface, riverine and estuarine input, and sediment seepage. Low molecular-weight hydrocarbons vary considerably both with season and area of the shelf. Higher surface-water methane values are apparent in the northern shallow waters and are probably related to the direct influence of riverine and estuarine factors. A unique higher occurrence in deeper waters on the southern extreme of the south Texas shelf is attributed to natural gas seepage across the mud-water interface. Other areas of high methane content are associated with the bottom water nepheloid layer, especially in summer. Microtarball concentrations in neuston samples are high in northern shelf waters and may be related to ship traffic at Aransas Pass Inlet and other points in the northern Gulf and to extensive petroleum activities in waters north of the south Texas shelf.

The lack of evidence of aromatic hydrocarbons in sediments suggests minimal petroleum pollution. Petroleum pollution in the form of microtarballs in the water column apparently does not contribute a sufficient quantity of petroleum hydrocarbons to the sediments. Concentrations of light hydrocarbons in the top few meters of shelf and slope sediments are highest near shore, decrease offshore, and are generally of microbial origin controlled by biological oxidation and diffusion into the overlying waters.

Studies of the effects of low-level and chronic input of petroleum in marine biota are complicated by the lack of information on hydrocarbon background levels in unpolluted environments, problems in differentiating petroleum compounds from biogenic hydrocarbons, and the effects of degradation of hydrocarbons, sediment absorption, interstitial water hydrocarbons, and the uptake from food. Approximately 50% of the zooplankton hydrocarbon samples in 1977 showed the possible presence of petroleum-like matter. This percentage was slightly more than that observed in 1976 (30%) and considerably higher than that in 1975 (7%). These values are higher than similar measures of particulate hydrocarbons in the water column, and they suggest that the majority of hydrocarbons in the zooplankton are not synthesized by them or by higher plants. The high values in zooplankton could be a reflection of bioaccumulation and concentrating tendencies of the environment near shore, characterized by seasonal fluctuations in suspended aluminosilicate particulate matter.

NANCY N. RABALAIS

ENVIRONMENTAL STUDIES OF A MARINE ECOSYSTEM
South Texas Outer Continental Shelf

1
INTRODUCTION
by R. W. Flint

The internal and external chemical, physical, and biological interactions of the world's oceans are among the most complex in the natural sciences. If the aspects and processes of these various interactions were understood, their scope and magnitude could be predicted for a given time and place. There are, however, many unknowns that must still be quantified.

The Texas coastal area is biologically and chemically a two-part marine system, the coastal estuaries and the broad continental shelf. These two components are separated by a chain of barrier islands and connected by inlets or passes. The area is rich in finfish and crustaceans, many of which are commercially and recreationally important. Many of the finfish and decapod crustaceans of this area exhibit a marine-estuarine dependent life cycle, that is spawning offshore, migrating shoreward as larvae and postlarvae, and utilizing the estuaries as nursery grounds (Gunter 1945; Galtsoff 1954; Copeland 1965). The broad continental shelf supports a valuable shrimp fishery that, as a living resource, contributes significantly to the local economy. Although an excellent overview of the zoogeography of this marine area is provided by Hedgpeth (1953), there are still many unknowns concerning the functioning of this complex system.

In 1974, the Bureau of Land Management (BLM), as the administrative agency responsible for leasing submerged federal lands, was authorized to initiate the National Outer Continental Shelf Environmental Studies Program. As part of this national program, the BLM developed the Marine Environmental Study Plan for the South Texas Outer Continental Shelf (STOCS) to add to our understanding of this particular ecosystem. This plan was developed to meet the following four specific study objectives:

[4] Introduction

1. Provide information for predicting the effects of outer continental shelf oil and gas development activities upon the components of the ecosystem;
2. Provide a description of the physical, chemical, geological, and biological components and their interactions against which subsequent changes or impacts could be measured;
3. Identify critical variables that should be incorporated into a monitoring program; and
4. Identify and conduct experimental and problem-oriented studies required to meet the basic objectives.

BLM contracted the University of Texas at Austin to act on behalf of a consortium program of research to be conducted by Rice University, Texas A&M University, and the University of Texas to implement this environmental study plan. The plan called for an intensive multidisciplinary three-year study (1975–1977) to characterize the temporal and spatial variation of the shelf marine ecosystem beyond the 10-m water depth.

In addition to the biological, physical, and chemical components reported herein, two other major field programs were conducted concurrently. The U.S. Geological Survey conducted a program designed to investigate suspended sediment flux, normal and storm transport and deposition of sediments, and sediment geochemistry in the STOCS area. The National Marine Fisheries Service of the National Oceanic and Atmospheric Administration conducted studies to investigate the historical distribution and abundance of ichthyoplankton in the area, to elucidate the snapper and grouper fisheries resources, and to determine the magnitude and economic significance of the recreational and associated "commercial/recreational" fisheries in the area. In addition to the above studies, which were restricted to the STOCS study area, Texas A&M University conducted a major field survey of the biological and chemical characteristics of selected topographic features in the northwestern Gulf.

The central purpose of the STOCS study was to provide an understanding of the living and nonliving resources of the shelf. An ecosystem is defined as "any area of nature that includes living organisms and non-living substances interacting to produce an exchange of material between the parts" (Odum 1959). In order to approach the objectives outline above, a broad program was designed that included

1. Water mass characterization;
2. Pelagic primary and secondary productivity as described by floral and faunal abundances, standing crop, and nutrient levels;

3. Sediment texture characterization;
4. Benthic productivity as described by infaunal and epifaunal densities;
5. Natural petroleum hydrocarbon levels in biota, water, and sediment; and
6. Natural trace metal levels in biota and particulate matter.

The additional final phase (1978–1979) of this study was devoted to the data synthesis and integration of the three previous years of sample collection and variable measurement. The goals of this synthesis and integration phase were twofold:

1. Develop a physical, chemical and biological description of the STOCS ecosystem, characterizing with a degree of confidence the temporal and spatial properties of those parameters that best represented the ecosystem between 1975 and 1977.
2. Develop mathematical descriptions for a few unique relationships defined by the data that will serve as "fingerprints" for future comparison by managerial decision makers and will contribute information to the general conceptual model.

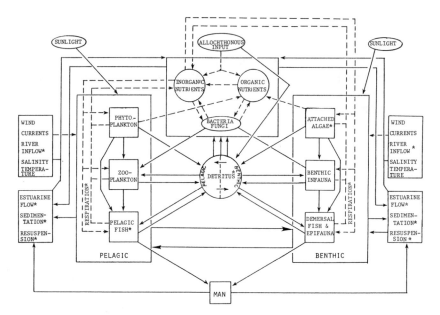

Figure 1. Conceptual model of the south Texas continental shelf ecosystem. Asterisk (*) indicates element not studied in STOCS investigation.

It was assumed that understanding the naturally inherent variability of this ecosystem would contribute immensely to evaluating potential impact of oil and gas exploration and production on the environment.

Using statistical techniques, we believed we could integrate the data base to the extent that an initial understanding of a typical marine ecosystem similar to the one depicted in Figure 1 could be documented. As shall be illustrated in the following chapters, in some cases we were relatively successful in developing an understanding concerning parts of this overall conceptual model, while in other cases, because of lack of sufficient information either within the data base or the supporting literature, we were not able to add detail to this model.

This book describes the STOCS research program and history. It integrates all study element results into a characterization of the STOCS ecosystem. Within the STOCS ecosystem there are many interrelated physical, chemical, and biological processes. We highlight some of these important factors and develop a conceptual model illustrating the manner in which they interact. We also include appendices (Appendix A and Appendix B) with statistics of all important study variables as identified either by the scientists in the study or by distinguishing spatial-temporal trends. The purpose of these appendices is to provide a quick reference concerning certain ecosystem variables along with their general statistical patterns to decision-makers and environmentalists assigned to the task of future monitoring. Additional details of the multidisciplinary program and contributions of the individual study areas are available in Parker (1976), Berryhill (1977), Groover (1977a), Griffin (1979), Flint and Griffin (1979), and Flint and Rabalais (1980).

Study Area

The general area of study corresponds to that portion of the Gulf of Mexico off the Texas coast designated by the U.S. Department of the Interior for future oil and gas leasing (Figure 2). The area covers approximately 19,250 km^2 and is bounded by 96° west longitude on the east, Pass Cavallo on the north, the Texas coastline on the west, and the Mexico–United States international border on the south. The Texas continental shelf has an average width of 88.5 km and a relatively gentle seaward gradient that averages 2.3 m/km.

No ecosystem is a completely self-contained unit, and the STOCS system is no exception. It is influenced by adjoining regions and waterways such as the open Gulf of Mexico, the Mississippi River to the northeast, the Rio Grande to the south, and the land masses to the

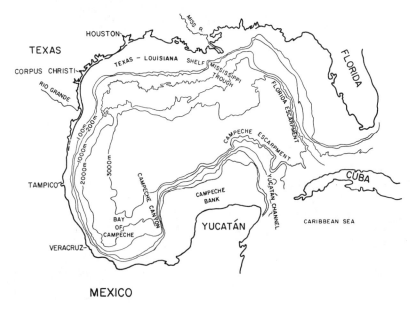

Figure 2. Bathymetric map of the Gulf of Mexico.

west. These adjacent bodies have a marked influence on the climate and are the sources of much input into the system. Although we can look at the region as a somewhat discrete unit, we must continually keep in mind these contiguous influences.

During the first year of study (1975), 12 sites corresponding to Stations 1–3 on four transects (I–IV) (Figure 3) were sampled. Thirteen additional transect sites were sampled during the second and third years of study, which included Stations 4–6 of Transects I–III and Stations 4–7 on Transect IV. For hydrographic studies, a seventh station was included on Transect II. These additional stations were added to increase shelf coverage of three special areas: (1) the shallow shelf environment (about 15 m deep) and its associated sandy sediments; (2) a zone in the middle of the study area that appeared anomalous in sediment characteristics, sediment trace metal content, and distributions of certain biological populations; and (3) a zone of active gas seepage near the shelf-slope break. In addition to the transect stations, four stations on each of the two submarine carbonate reefs, Hospital Rock (HR) and Southern Bank (SB), were sampled in 1976. Collections were decreased to two stations at each reef in 1977. Table 1 lists the long-range navigation (LORAN) and long-range accuracy

[8] Introduction

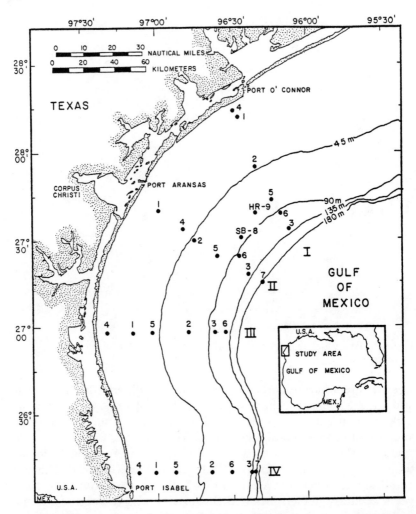

Figure 3. Map of the south Texas continental shelf bathymetry and location of sampling sites. Insert shows location of study area in relation to the entire Gulf of Mexico.

Introduction [9]

Table 1. Monitoring Study Station Locations and Depths

Tran.	Sta.	LORAN 3H3	3H2	LORAC LG	LR	Latitude	Longitude	Depth Meters	Feet
I	1	2575	4003	1180.07	171.46	28°12′N	96°27′W	18	59
	2	2440	3950	961.49	275.71	27°55′N	96°20′W	42	138
	3	2300	3863	799.45	466.07	27°34′N	96°07′W	134	439
	4	2583	4015	1206.53	157.92	28°14′N	96°29′W	10	33
	5	2360	3910	861.09	369.08	27°44′N	96°14′W	82	269
	6	2330	3892	819.72	412.96	27°39′N	96°12′W	100	328
II	1	2078	3962	373.62	192.04	27°40′N	96°59′W	22	72
	2	2050	3918	454.46	382.00	27°30′N	96°45′W	49	161
	3	2040	3850	564.67	585.52	27°18′N	96°23′W	131	430
	4	2058	3936	431.26	310.30	27°34′N	96°50′W	36	112
	5	2032	3992	498.85	487.62	27°24′N	96°36′W	78	256
	6	2068	3878	560.54	506.34	27°24′N	96°29′W	98	322
	7	2045	3835			27°15′N	96°18.5′W	182	600
III	1	1585	3880	139.13	909.98	26°58′N	97°11′W	25	82
	2	1683	3841	286.38	855.91	26°58′N	96°48′W	65	213
	3	1775	3812	391.06	829.02	26°58′N	96°33′W	106	348
	4	1552	3885	95.64	928.13	26°58′N	97°20′W	15	49
	5	1623	3867	192.19	888.06	26°58′N	97°02′W	40	131
	6	1790	3808	411.48	824.57	26°58′N	96°30′W	125	410
IV	1	1130	3747	187.50	1423.50	26°10′N	97°01′W	27	88
	2	1300	3700	271.99	1310.61	26°10′N	96°39′W	47	154
	3	1425	3663	333.77	1241.34	26°10′N	96°24′W	91	298
	4	1073	3763	163.42	1456.90	26°10′N	97°08′W	15	49
	5	1170	3738	213.13	1387.45	26°10′N	96°54′W	37	121
	6	1355	3685	304.76	1272.48	26°10′N	96°31′W	65	213
	7	1448	3659	350.37	1224.51	26°10′N	96°20′W	130	426
HR	1	2159	3900	635.06	422.83	27°32′05″	96°28′19″	75	246
	2	2169	3902	644.54	416.95	27°32′46″	96°27′25″	72	237
	3	2163	3900	641.60	425.10	27°32′05″	96°27′35″	81	266
	4	2165	3905	638.40	411.18	27°33′02″	96°29′03″	76	250
SB	1	2086	3889	563.00	468.28	27°26′49″	96°31′18″	81	266
	2	2081	3889	560.95	475.80	27°26′14″	96°31′02″	82	269
	3	2074	3890	552.92	475.15	27°26′06″	96°31′47″	82	269
	4	2078	3890	551.12	472.73	27°26′14″	96°32′07″	82	269

(LORAC) coordinates, latitude, longitude, and depth of each site of sampling during the three-year study.

Sampling and Program Description

The field investigations for the first year of the study began in early December 1974 and were completed by mid-October 1975. Samples were collected from Stations 1–3 of all transects for the first year of study during three biological-meteorological seasons. The three seasons were winter (December–February), spring (April–May), and fall (September–October). For more exact cruise dates, see Parker (1976).

The field sampling for the second year of study was initiated mid-January 1976 and was completed mid-December 1976. Samples were collected during three biological-meteorological seasons from all transects and the bank stations. The three seasons were winter (January–February), spring (May–June), and fall (September–October). In addition to the seasonal sampling, samples were collected from Transect II and the bank stations in the six months (March, April, July, August, November, December) not included in the three seasonal sampling periods. For more exact cruise dates, see Groover (1977a).

Basing their decision on the initial results from 1976, both BLM and the scientists determined that additional information on certain study elements was needed to meet the objectives of the investigation. Consequently, several supplemental studies were initiated in September 1977 and completed in November 1978. The results of these studies were reported to BLM in January 1979 (Griffin 1979). In addition to these, a separate rig monitoring study was also initiated in late 1976. The objectives of that study, which included characterizing the effects of drilling muds, cuttings, and other disposals associated with exploratory drilling, were met by surveys of the sediments, organisms, and water in the immediate vicinity of an exploratory drilling rig before, during, and after the drilling (Groover 1977b).

The field sampling for the third and final year of study was initiated mid-January 1977 and was completed mid-December 1977. Samples were collected during three biological-meteorological seasons from all transects and four of the bank stations. The three seasons were winter (January–February), spring (May–June), and fall (September–October). In addition to the seasonal samplings, samples of some of the study elements were collected from Transect II during the six months (March, April, July, August, November, and December)

Table 2. STOCS Biological, Chemical, and Physical Components Study: Participants by Work Element and Institution

Rice University	
Microplankton and shelled microzoobenthon	R. E. Casey
Texas A&M University	
High molecular-weight hydrocarbons in macroepifauna, demersal fish, and macronekton	C. S. Giam, H. S. Chan, G. Neff
Trace metals in macroepifauna, demersal fish, macronekton, and plankton	B. J. Presley, P. N. Boothe
Low molecular-weight hydrocarbons, nutrients, and dissolved oxygen	W. M. Sackett, J. M. Brooks, B. B. Bernard
Zooplankton	E. T. Park, P. Turk
Neuston	J. H. Wormuth, L. H. Pequegnat, J. D. McEachran
Meiofauna	W. E. Pequegnat, C. Venn
Histopathology of macroepifauna	J. M. Neff, V. Ernst
Histopathology of demersal fishes	W. E. Haensly, J. Eurell
Benthic bacteriology	J. R. Schwarz, S. K. Alexander
The University of Texas	
AUSTIN	
Water column and benthic mycology	P. J. Szaniszlo, P. Powell
MARINE SCIENCE INSTITUTE/GALVESTON	
GEOPHYSICAL LABORATORY	
Sediment texture	E. W. Behrens
MARINE SCIENCE INSTITUTE/PORT ARANSAS MARINE LABORATORY	
Ciliated protozoa	P. L. Johansen
Hydrography	N. P. Smith
High molecular-weight hydrocarbons in zooplankton, sediment, water	P. L. Parker, R. S. Scalan, J. K. Winters
Phytoplankton and productivity	C. Van Baalen, D. L. Kamykowski, W. Pulich
Macroinfauna and macroepifauna	J. S. Holland
Demersal fish	D. E. Wohlschlag, R. Yoshiyama
SAN ANTONIO	
Histopathology: Gonadal tissues of macroepifauna and demersal fish	S. A. Ramirez
Water column bacteriology	M. N. Guentzel, H. V. Oujesky, O. W. Van Auken

Table 3. List of Study Areas and Environmental Variables Focused upon During the Data Synthesis and Integration Aspect

	Study Area	*Variables*
	Hydrography	Temperature
		Salinity
		Depth
		Currents
		Secchi depth
		Transmission
	Nutrients	Silicate
		Phosphate
Pelagic		Nitrate
nonliving		Dissolved oxygen
characteristics	Hydrocarbons	
	Low molecular-weight (LMW)	Methane
		Ethane
		Ethene
		Propene
		Propane
	High molecular-weight (HMW)	Hexane or benzene fractions
		Retention Index w/concentrations
	Phytoplankton	Species densities
		Chlorophyll (biomass)
		^{14}C productivity
		ATP
	Microorganisms (bacteriology & mycology)	Species abundances
Pelagic		Total counts and hydrocarbonoclastic counts
living		
characteristics	Neuston	Species densities
		Tarball concentrations
	Zooplankton (micro & macro)	Species densities
		Sample biomass
		Trace metal body burden
		HMW hydrocarbon body burden

	Study Area	Variables
	Sediment texture	Mean grain size Percent sand Percent silt Percent clay

Benthic
nonliving
characteristics
	Sediment chemistry	Organic carbon
		Delta ^{13}C
		Ethene
		Ethane
		Propene
		Propane
		Methane
		HMW hydrocarbons
		   Hexane or benzene fractions
		   Retention Index w/concentrations

	Microorganisms (bacteriology & mycology)	Species abundances Total counts and hydrocarbonoclastic counts
	Meiofauna	Species densities
Benthic living characteristics	Macroinfauna	Species densities
	Invertebrate Macroepifauna	Species densities Trace metal body burdens HMW hydrocarbon body burdens Tissue histopathology
	Demersal fish	Species densities Biomass Trace metal body burdens HMW hydrocarbon body burdens Tissue histopathology

not included in the three seasonal sampling periods. For more exact cruise dates, see Flint and Griffin (1979).

All sample collections and measurements, except the placement and recovery of current meters, were taken aboard the University of Texas research vessel, the *Longhorn*. The *Longhorn*, designed and constructed as a coastal research vessel in 1971, is a steel-hulled, 24.38 m (80 ft) by 7.32 m (24 ft) by 2.13 m (7 ft) draft ship. She carries a crew of five and can accommodate a scientific party of ten. The *Longhorn* is equipped with a stern-mounted crane, a trawling winch, side-scan sonar, radar, LORAN-A and LORAC navigational systems, and dry and wet laboratory space. Navigation and station location for water column cruises were by LORAN. Navigation and station location for benthic cruises were by LORAC navigational systems.

The University of Texas Marine Science Institute's Port Aransas Marine Laboratory provided overall project management, logistics, ship time, data management, and certain scientific services for the study. Additional scientific effort was provided by Texas A&M University, the University of Texas at San Antonio, the University of Texas at Austin, and Rice University.

A total of 28 principal investigators participated in the three-year sampling program. Table 2 lists the principal investigators and their associates with their respective institutions and scientific responsibilities. For the final year of the program, during the data synthesis and integration effort, BLM emphasized fewer study elements than the 20 listed in Table 2. The specific study areas and variables considered in the synthesis and integration effort are listed in Table 3. Researchers anticipated that the analysis design developed would provide knowledge of the various living and nonliving components in sufficient detail to begin to understand their relationships and enhance the ability to anticipate changes resulting from pollution of the STOCS ecosystem. Complete descriptions of sampling methods and laboratory analyses are included in Flint and Rabalais (1980).

2
MARINE PELAGIC ENVIRONMENT

with contributions by J. M. BROOKS, D. L. KAMYKOWSKI, P. L. PARKER, R. S. SCALAN, N. P. SMITH, J. K. WINTERS

Marine Meteorology

Patterns and trends in the marine environment of the northwestern Gulf of Mexico are strongly influenced by various kinds of meteorological events. Water levels along the coast change noticeably with changes in wind speed and direction. Typical circumstances for these water level changes are associated with hurricanes and the quick changes in wind directions from "northers," winter's high pressure waves. The stress of the wind acting upon the sea surface at times other than hurricanes and "northers," however, may also be sufficient to bring about water level changes of the same magnitude as those resulting from a periodic tide-producing force. This stress may lead to considerable deviation from the observed water level changes published in tide manuals and may help explain the extremely unpredictable nature of water level changes along the Gulf coast.

In turn, numerous characteristics of the Gulf surface waters influence many of the weather patterns observed in the northwestern Gulf of Mexico. On a large scale, for example, the relatively high temperature of Gulf surface waters, compared with that of other waters in the same latitudes, brings about such a great warming and an increase in the moisture content of the overlying air masses that weather patterns of the northwestern Gulf are markedly affected. A discussion of the extent to which the sea surface affects the overlying atmosphere may be found in Jacobs (1951). He computes the average winter evaporation in the Gulf to be approximately 0.4 g/cm^2/day and compares this with the average evaporation of other ocean areas of the world. As one would expect, there is a strong coupling of atmospheric conditions and the sea surface conditions in the Gulf of Mexico, conditions that serve as driving forces affecting the dynamics of the marine environment.

The climate of south Texas is subtropical and is characterized by short, mild winters and hot summers. Significant variations in this trend do occur from north to south along the coastline. The climate of the coastal plain from the Texas-Louisiana border to Corpus Christi can be characterized as subhumid, with the area from Matagorda Bay to Corpus Christi considered more dry and subhumid than the area from the Texas-Louisiana border to Matagorda Bay (Hedgpeth 1953). Rainfall along the Texas coast averages 25 to 125 cm/yr but decreases significantly closer to Corpus Christi. Compare Galveston's average rainfall of 106.2 cm/yr to Corpus Christi's 71.9 cm/yr. Rivers in the Corpus Christi area are small and contribute much less freshwater to the estuaries than those farther north. Air temperatures in the south Texas area are higher in the summer than those along the Louisiana coast, and in the winter this area may have the lowest temperatures observed for the entire Gulf coast (Parker 1960). In this dry, subhumid portion of the coast, salinities of estuaries and lagoons commonly vary from medium to high as heavy rainfall increases riverine input or as evaporation exceeds runoff for extensive time periods.

In contrast, the coastal zone from Corpus Christi to the mouth of the Rio Grande is classed as semiarid. No permanent rivers flow into the lagoons and estuaries, resulting in continuous hypersalinity. Rainfall is frequently less than 30 to 70 cm/yr, and summer air temperatures can exceed 42°C. In comparison to Corpus Christi, Brownsville receives an average rainfall of 67.9 cm/yr, an average 4 cm/yr less.

Average sea level atmospheric pressures in the Gulf of Mexico vary from 76.2 to 79.1 cmHg. There are wide deviations from these pressures in individual synoptic circumstances such as during the development of tropical storms. Superimposed upon these general annual patterns are diel pressure variations.

During a typical 24-h period there is an early morning minimum in atmospheric pressure followed by a late morning maximum, an evening minimum greater than that of the morning and a nocturnal maximum less than that of the morning (Leipper 1954). The average winds vary from 6 to 8 knots (11.1 to 14.8 km/hr) in the summer, with stronger more variable winds from 10 to 12 knots (18.5 to 22.2 km/hr) in the winter. Fog is most frequent in midwinter and occurs most often in the north central part of the Gulf. For the annual period, the average cloud cover over the northwestern Gulf of Mexico ranges from 40% to 60% of the sky obscured. The low cloud type most commonly reported is cumulus (Leipper 1954).

The general circulation of air near the Gulf surface over the south Texas coastal region follows the sweep of the western extension of the Bermuda high-pressure system throughout the year. The Bermuda pressure system becomes dominant during the spring months, as the influence of northern anticyclones (low-pressure areas) causing northerly fronts disappears. Mean barometric pressure falls, and the minimum mean pressure occurs in the summer as the equatorial trough migrates northward, allowing prevailing southeasterly winds to dominate. At this point, the low-pressure systems over Mexico deepen significantly.

Beginning in September, the equatorial trough moves southward, the Mexican low-pressure system fills, and the Bermuda high-pressure system decreases in strength. Accompanying this trend, continental high-pressure systems to the north intensify as winter approaches. As barriers weaken to the south, the high-pressure systems moving from the north reach the lower latitudes and produce maximum barometric pressures in the winter. The result of these conditions is an increase in the frequency of "northers" over the south Texas coast. The high-pressure systems and their associated extratropical cyclones are responsible for the wide pressure ranges observed in winter (Berryhill 1977).

Air temperature extremes for the south Texas area are influenced primarily by the combined effects of prevailing southeasterly winds and the large expanse of Gulf waters. Low temperatures occur when strong northerly winds associated with cold fronts penetrate the area. Freezing temperatures normally occur in coastal areas at least once each winter. The highest summer temperatures occur when the wind direction shifts from the prevailing southeast to south and southwest.

As mentioned previously, south Texas is semiarid. Peak precipitation months are May and September. Tropical cyclones may add large amounts to monthly rainfall totals from June to October and may cause bays normally high in salinity to freshen drastically in only a few hours. Winter months have the least rainfall. Winter precipitation comes mainly from frontal activity and low stratus clouds. Because of the semiarid conditions, not only along the coast but landward for more than 100 miles, no major streams flow to the Gulf of Mexico along the south Texas coast between Aransas Pass Inlet and the Rio Grande, 135 miles to the south. This factor has a direct influence on the pattern of marine processes on the south Texas shelf.

Compared to those of the adjacent land area, offshore winter temperatures of the south Texas shelf are higher and average wind velocities are greater. Offshore summer conditions are more similar to the onshore climate but with some diurnal differences: the daily temperature range is narrower and the maximum afternoon wind speed is lower offshore than at stations along the coast. The offshore area, unlike the coastal area, does not exhibit a season of extensive rainfall. Rain is most frequent in December and January with a secondary peak in August and September related to tropical storms and depressions. Based on rain frequency, the driest season in the offshore area is March to June, with an average of less than 3% of the ship's weather observations reporting rain. When rainfall was recorded in the cruise reports, it was most frequently noted as occurring midafternoon.

Physical Oceanography

Hydrographic features illustrate the annual progression that can occur over a shelf area such as that of the south Texas Gulf of Mexico. These descriptions can also provide insight concerning possible factors that influence the functioning of the ecosystem. Taken together, a number of previous studies including Jones, Copeland, and Hoese (1965), Rivas (1969), Armstrong (1976) and Devine (1976) provide a good overview of the hydrographic conditions of Texas shelf waters.

In the surface layer, strong cross-shelf temperature gradients during midwinter months disappear with seasonal heating, and surface water becomes spatially isothermal at approximately 29°C by late summer. Vertical stratification, on the other hand, is nearly absent in shelf waters during the winter months, but it is well developed in summer. Shelf salinities remain relatively high for most of the year. An exception is a short period during spring and early summer when a plume of Mississippi River water may cover the entire shelf, lowering salinities through the uppermost 20 to 30 m.

In the open waters of the Gulf of Mexico, water mass distributions are the result of inflow through the Yucatán Channel, outflow through the Straits of Florida, surface conditioning by local air-sea exchange processes, and internal mixing. Together, these influences produce three well-defined water masses in layers below the mixed layer of the surface. The sill depth of approximately 2000 m between the Yucatán Peninsula of Mexico and the western tip of Cuba exerts significant influence on the temperature and salinity distribution in open Gulf waters. Below the sill depth, both temperature and salinity are characterized by spatial homogeneity due to the isolation from the deep-to-bottom water found in the Atlantic Ocean and the Caribbean Sea. Gulf basin water, found below the effective sill depth of approximately 1500 m, is characterized by potential temperatures between approximately 4.2°C and 4.4°C, and salinities between 34.96‰ and 34.98‰.

Above the Gulf basin water, salinities decrease to 34.86‰ between depths of 900 to 1100 m. This salinity minimum reflects the influence of the Antarctic intermediate water, which can be traced back through the Caribbean Sea, across the tropical Atlantic Ocean, and into high southern latitudes to a source at the Atlantic polar front at 45° to 50° south latitude.

Both temperature and salinity increase as depth decreases above the layer of the Antarctic intermediate water. A maximum in salinity is characteristically found between approximately 100 and 300 m. This feature can also be traced back through the Caribbean Sea and upstream along the subtropical undercurrent to a source under the semipermanent high-pressure center in the Atlantic Ocean east of Bermuda (Wüst 1964, Nowlin 1971). Salinities are generally between 36.2‰ and 36.7‰, while temperatures characteristic of this layer vary between about 18°C and 26°C in open Gulf waters.

The surface mixed layer lies on top of the three distinct water masses discussed above. Because of the direct contact this layer has with the overlying atmosphere, it is often quickly modified or conditioned by conductive, evaporative, and radiative processes. Thus, the shelf waters exhibit not only a well-defined annual cycle but also substantial variability over shorter time scales.

Temperatures characteristic of the mixed layer over the inner Texas shelf range from approximately 11°C to 13°C in late winter to 28°C to 29°C in late summer. Salinity in waters near shore is also variable, ranging from open Gulf surface values of approximately 36.4‰ to 20‰ or less during the spring run-off or periods of heavy rainfall.

The extensive study and observation of south Texas shelf waters for three years has provided information that supports many of the patterns outlined above. In addition, many of the observations have significantly refined our understanding of the physical dynamics of the complex Texas shelf waters. A good hydrographic summary of these waters can be obtained by examining time-depth plots of temperature and salinity for a shallow and a deep station on the shelf (Figure 4). At the deeper station, salinity varies minimally with the exception of lower surface salinities in the spring of the year. Temperatures indicate a greater degree of variability, but there is no well-defined pattern at depth with the exception of a prevalent stratification during the summer of each year. Surface temperatures suggest a sinusoidal variation with highest temperatures occurring in August.

Hydrographic data from surface and bottom layers at the shallow

[20] Marine Pelagic Environment

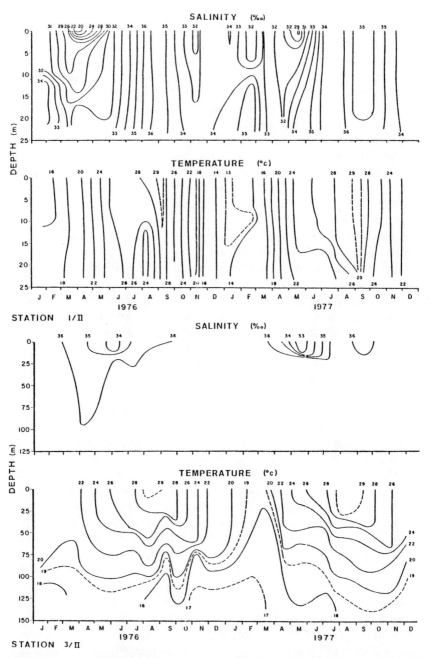

Figure 4. Time-depth plots of water temperature and salinity from Transect II inner (Station 1) and outer (Station 3) shelf stations for January through December 1976 and 1977.

site show a much greater vertical homogeneity and a more clearly defined seasonal variation at both depths. The water column over the inner shelf is nearly isothermal during the fall, winter, and spring months. During midsummer a slight stratification is sometimes present. Salinities at this site are almost totally influenced by local rainfall and riverine input.

A comparison of surface temperature at these two sites provides a rough picture of cross-shelf dynamics over the annual cycle. During the summer months, temperatures of slightly over 29°C are observed at both stations, suggesting minimal cross-shelf gradients. In contrast, lowest values (approximately 14°C) over the inner shelf are well below the minimum values of 19°C to 20°C found over the outer shelf during the winter, resulting in strong cross-shelf gradients during these months.

Figure 5 illustrates another aspect of the hydrography that is of prime importance to many of the biological communities, especially benthic populations, that of variation in the hydrographic environment over time. There is a significant negative correlation ($P < 0.05$) between bottom water temperature and salinity standard deviations and depth, indicating the extreme variability of shallow waters and contrasting stability of deeper waters. A deviation from this trend is noted for several of the collection sites deeper than 100 m. Increases

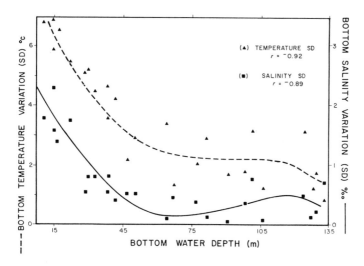

Figure 5. Plot of water temperature and salinity standard deviation (SD) calculated over the study duration against water depth of each station where samples were taken. The r values of the two polynomial curves are shown.

in the variance of salinity at these sites may suggest the occurrence of occasional upwelling of deep Gulf waters. This possibility is further verified by a plot of a water temperature cross-section along a transect during the summer of 1977 (Figure 6). Warmest waters were found in surface layers at some distance from the coast. The onshore directed temperature gradient together with the layer of cool near-bottom water extending nearly to the coast indicate the existence of upwelling and a pattern of offshore Ekman transport of surface water with a near-bottom return flow. The summer isothermal conditions across the shelf are ideal for this phenomenon to occur and are the only opportunity for cross-shelf currents perpendicular to the coast to occur with any regularity because of the predominant wind directions from the south-southeast.

The graphical summaries of the hydrographic data presented here are useful for quantifying the spatial as well as the temporal variability in the hydrographic climate in Texas shelf waters. The time scales associated with the dominant local variations in temperature and salinity differ significantly between inner- and outer-shelf sites. There was a lack of stratification over the inner shelf at all depths as well as through the surface layers of the entire shelf during most of the summer months.

At greater depths, sufficiently removed from surface conditioning

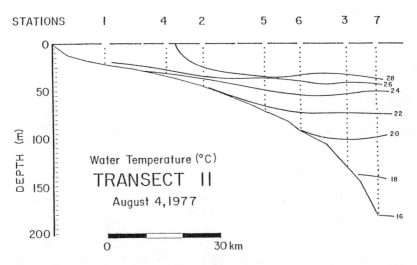

Figure 6. Water temperature cross-section along Transect II, 4 August 1977.

by air-sea exchange processes, the dominant time scales become too short to be properly resolved with the available data. If the temperature variations recorded at near-bottom levels at the outer station are associated with a vertical movement of the top of the permanent thermocline, the associated time scales would be on the order of an hour to several days. These scales would depend on whether these variations reflect internal waves or a meteorologically forced encroachment of water onto the shelf from the open Gulf.

There appear to be several significant events recurring annually and making a well-defined impression in the hydrographic record of the south Texas shelf. The plume of Mississippi River water, moving westward and southwestward along the northern rim of the Gulf of Mexico during the winter and spring months, is especially pronounced near the coast but may at times cover the entire shelf. In addition, local rivers and estuaries have potential to influence parts of the shelf, especially coastal waters, during parts of the year.

Examination of trends in a phytoplankton biomass indicator, chlorophyll *a*, provided additional evidence over the study period of different water mass influences on the Texas shelf and refined ideas concerning general physical dynamics of the ecosystem. Of the various processes contributing to the variability of plant biomass across the shelf, freshwater discharge appeared to be most influential of those variables examined during the study. Figure 7 illustrates the relationship between salinity and particulate matter in the water column. This relationship suggested that as salinity decreases from riverine input the particulate matter increases (decreased secchi depth) along with possible associated nutrients and primary productivity. The collection of surface water samples along Transect II (off the Aransas Pass Inlet to approximately 90 km offshore) provided an ideal picture of the relationship between waters low in salinity, which are potentially related to freshwater inflow, and chlorophyll *a* concentrations (Figure 8). The highest monthly concentrations of chlorophyll *a* were usually associated with lower salinities, usually less than 30‰. This was especially apparent in the offshore waters in late winter and early spring. In contrast, the variations in temperature did not appear to play an influential role in chlorophyll trends.

Through correlational research utilizing salinity patterns, chlorophyll *a* concentrations, and river flow values, we demonstrated that the STOCS area may be influenced by different freshwater sources, depending upon distance from shore on the shelf. Figure 9 summarizes the relationships among chlorophyll *a*, salinity, and freshwater inflow from five point sources hypothesized as influencing the STOCS

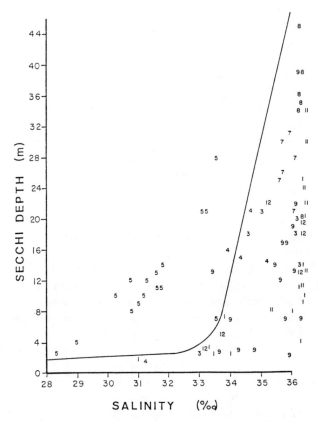

Figure 7. Relationship between salinity and secchi depth for all stations. Solid line is arbitrary curve. Numbers correspond to months of year.

area. The top part of the figure is a plot of correlation coefficients against distance offshore (km). The correlation coefficients interrelate the 12 chlorophyll *a* and salinity values available for successive 1.85-km distances offshore. The zones (marked by vertical lines) within this plot are based on the results of plotting similar correlation coefficients against distance offshore, interrelating point source discharge with either salinity or chlorophyll *a*. The zones of maximum negative correlation with salinity (dashes) or of maximum positive correlation with chlorophyll *a* (dots) are shown for each point source in the bottom part of Figure 9.

An inshore zone between 0 and 14 km offshore is characterized by a high average correlation (−0.76) between chlorophyll *a* and salinity

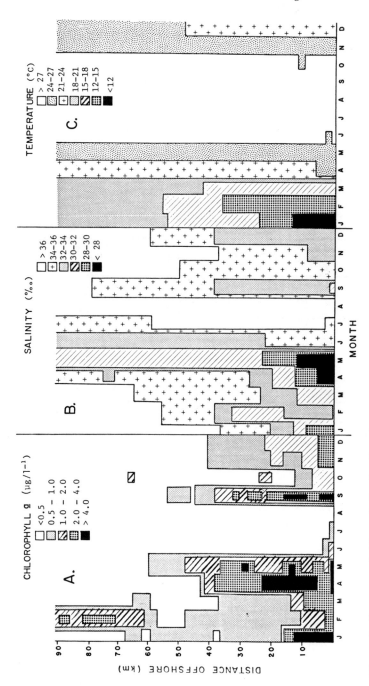

Figure 8. Contours of chlorophyll *a* (μg/l), salinity (‰), and temperature (°C) based on collections made every month in 1977 at points 1.85 km apart along a transect extending from the Port Aransas jetties to Station 3, Transect II.

[26] Marine Pelagic Environment

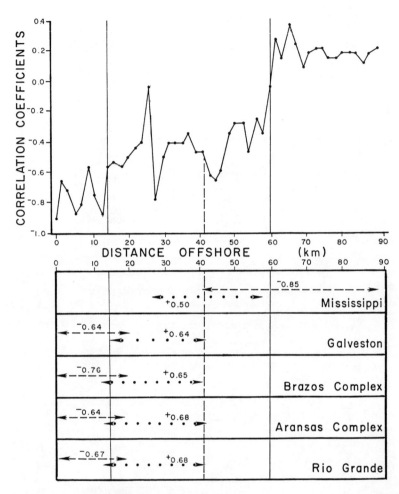

Figure 9. Relationships between chlorophyll *a*, salinity, and freshwater sources. *Top*, plot of distance offshore (km) against the correlation coefficients of monthly chlorophyll *a* and salinity (12 points) for successive distances offshore. *Bottom*, the zones of maximum correlation between monthly point source discharge and either monthly salinity (dashes) or monthly chlorophyll *a* (dots) for five significant freshwater sources in the northwest Gulf of Mexico. For example, monthly Mississippi River discharge in 1977 exhibited an average correlation of −0.85 and monthly salinity readings for every 1.85 km between 41 and 90 km offshore.

and by the highest correlations between Texas point source discharge and salinity. Chlorophyll *a* is not well correlated with any point source discharge within this zone.

The middle zone extends from 14 to 59 km offshore. The average chlorophyll *a*–salinity correlation (−0.41) decreases in this region. Neither the Texas source discharges nor the Mississippi River discharge is well correlated with salinity throughout this zone. Texas river discharge, however, is related to salinity at the inshore side of the zone, and Mississippi River discharge is highly related to salinity at the offshore side of this zone. The major correlations between point source discharge and chlorophyll *a* almost exclusively occur in this zone. The point sources north of the sampling transect yield an interesting pattern: the farther away the point source, the farther offshore occurs the band of highest correlation. The Rio Grande exhibits its highest correlation with chlorophyll *a* between 39 and 50 km offshore. The Texas point sources to the north of the cross-shelf transect all abruptly end their high correlation with chlorophyll *a* at 41 km offshore. This feature divides the middle zone into two subzones: between 14 and 41 km offshore, chlorophyll *a* is best related to Texas freshwater sources; between 41 and 59 km offshore, chlorophyll *a* is best related to Mississippi River discharge.

The offshore zone extends from 59 km offshore to the end of the transect (90 km). The average chlorophyll *a*–salinity correlation (+0.21) becomes positive in this region, suggesting fresh water does not contribute to increased chlorophyll *a*. In fact, chlorophyll *a* shows a tendency to decrease with decreasing salinity. Mississippi River discharge is highly correlated with salinity in this zone.

The preceding description provides sufficient information to develop the following model to aid in explaining the potential water mass dynamics on the Texas shelf.

1. Inshore Zone: The dominant force is fresh water from Texas riverine input. Salinity is inversely correlated with river discharge, but chlorophyll *a* shows no pattern. Although chlorophyll is highly correlated with salinity, phytoplankton patchiness due to other factors (e.g., sediment resuspension or grazing) in this shallow, well-mixed area apparently confounds a consistent relationship between chlorophyll and river discharge.
2. Middle Zone: Texas freshwater sources as well as the Mississippi River influence both salinity and chlorophyll *a*. Because of mixing, freshwater discharge from the different sources shows its strongest relationship with salinity at the zone boundaries. The correla-

[28]   Marine Pelagic Environment

tions inshore of 41 km suggest a strong Texas freshwater presence, while correlations beyond indicate the Mississippi River is the more significant freshwater source.

3. Offshore Zone: Mississippi River discharge dominates the shelf beyond approximately 59 km offshore. The salinity-point source discharge correlation is highly negative. The point source discharge correlation with chlorophyll, however, is weak and also shows a negative response. This seems intuitively proper, since the transit time from the source to the STOCS area is probably sufficient to deplete nutrients.

From the patterns described above it is possible to understand more easily the gradients that exist on the Texas shelf and why they exist in moving from coastal shallow waters with local influences to deeper more ocean-like waters farther out on the shelf that have very distant influential processes driving their dynamics. The conceptual model developed above suggests that many of the dynamics of the Texas shelf, such as those associated with pelagic biota, can be explained by considering topography, local river input, Mississippi River discharge, and climatic variables such as wind direction and velocity.

Between approximately October and March, the currents along the shelf at Aransas Pass Inlet are toward the south-southwest with a predominant longshore component. Between June and September, currents over the Texas shelf are substantially weaker. The longshore component reverses over very short time scales, and there are often periods of water movement across the shelf perpendicular to the coast as described above.

Drift bottle observations from a separate study (Watson and Behrens 1970) indicated that most of the currents directly off the barrier islands of the Texas coast were generated by local winds as measured at Corpus Christi. Currents near shore were observed flowing in opposite directions during winter and summer, direction correlated to the prevailing winds. During periods of transitional weather, especially in the spring, southward surface drift was often counter to local southerly winds. Apparently the Texas coastal waters are affected by significant currents generated by winds representative of winter conditions in another, probably more northern part of the Gulf, while summer winds have begun to blow in the south Texas region.

The seasonal variation in shelf circulation has a direct and obvious effect on the spatial distribution and temporal variability of hydrographic variables and suggests possible influential factors affecting ecosystem dynamics. The strong and quasi-steady water flow to the south-southwest during the winter months, and especially into late spring, is responsible for the advective transport of Mississippi River water along the northwestern rim of the Gulf of Mexico at a time when discharge is at its maximum. During the summer months, aperiodic near-bottom encroachment of water from the depths of the outer shelf may play an important role in ecosystem dynamics during times of relatively low riverine input. The importance of cross-shelf motion in transporting nutrients, heat, suspended solids, live plankton, or any combination of these factors becomes quite apparent.

Water Chemistry

NUTRIENTS Nutrient concentrations of the Gulf waters observed during the study were representative of open Gulf surface waters in most of the water above 60 m depth, but as illustrated in Figure 7, continental run-off influenced nearshore surface concentrations, especially in the spring. Nitrate, as the limiting nutrient to primary production, decreased to concentrations essentially below detection limits (< 0.1 μM/l) after the spring and early summer phytoplankton bloomed each year (Figure 10). Phosphate and silicate were substantially affected by the blooms each year but were never completely depleted during summer periods. These nutrients were generally replenished in the water column during the fall, reaching their maxima in the early winter to midwinter period. Contrasts between shallow and deep sites on the shelf for surface water concentrations of these nutrients generally illustrated similar trends, with the station more distant from the coastline showing proportionately lower concentrations. The exception to these patterns was observed for phosphate; the inshore station appeared to show a completely different trend than the deeper, offshore station. The intrusion of nutrient-rich western Gulf waters usually found between 200 and 300 m was often seen below 70 m as indicated by the phosphate concentration cross-shelf contours from Transect II (Figure 11).

Oxygen concentrations in the upper 60 m of water varied seasonally, being generally highest in the winter and lowest in the summer (Figure 10). The two sites again illustrated similar trends for surface water concentrations. Ratios of measured oxygen to equilibrium oxygen concentrations indicated that oxygen variations in the upper 60 m were generally controlled by physical processes (seasonal hydrographic variability) rather than pelagic productivity fluctuations. Masses of highly oxygenated water could be traced by cross-sectional contours as they were formed near shore in the water and displaced by warming in the spring and summer. The intrusion of oxygen-poor

[30] Marine Pelagic Environment

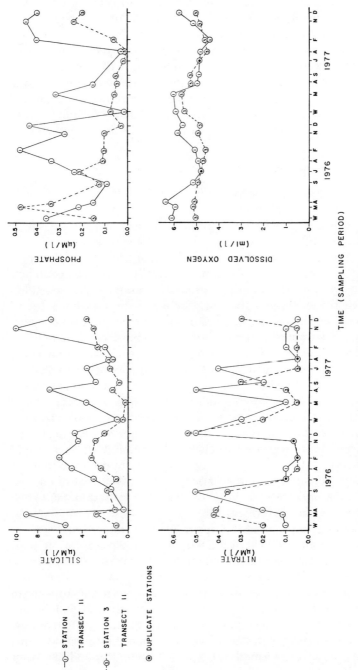

Figure 10. Silicate (μM/l), phosphate (μM/l), nitrate (μM/l), and dissolved oxygen (ml/l) for the surface waters at Stations 1 and 3 of Transect II. Samples were collected in winter (W), March (M), April (A), spring (S), July (J), August (A), fall (F), November (N), and December (D) in 1976 and 1977.

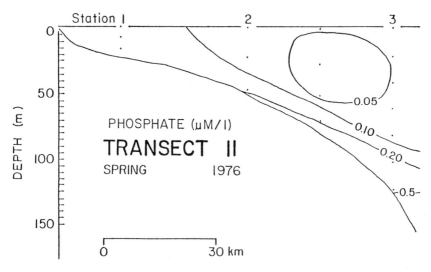

Figure 11. Phosphate (μM/l) cross-sectional contours along Transect II during the spring seasonal sampling (1976).

central Gulf water usually found between 200 and 300 m was often evident below approximately 70 m water depth. Seasonal variation of oxygen concentrations through the water column could be seen and were related to the vertical extent of mixing in deeper offshore waters. In general, bottom water oxygen concentrations observed during this study (2.58–5.86 ml/l) appeared to be sufficient to support organisms on the Gulf sea floor.

HYDROCARBONS   The STOCS ecosystem is relatively pristine concerning hydrocarbons, and those observed during this study we attributed primarily to natural sources. Numerous investigators have established that the open ocean is an important source of methane to the atmosphere. Much of the methane that fluxes at the air-water interface can be attributed to in situ production associated with water levels highly active in primary production and possibly bacterial production within the water column. Other natural sources of methane and other low molecular-weight hydrocarbons (LMWH) include riverine and estuarine input as well as sediment seepage.

Methane in the northwestern Gulf water column exhibited considerable seasonal and spatial variability during the study period as indicated by the ranges in Table 4. Higher surface methane values were

Table 4. Number of Observations, Mean, Minimum, and Maximum Low Molecular-Weight Hydrocarbon Concentrations at Stations During 1976 and 1977

Hydrocarbon	Observations	1976 Mean (nl/l)	(Min.–Max.) (nl/l)
Methane	299	97	(41–500)
Surface methane	54	73	(41–157)
Ethene	219	4.5	(0.1–25)
Surface ethene	54	6.7	(0.2–25)
Ethane	108	0.4	(0.1–1.3)
Surface ethane	53	0.4	(0.1–0.9)
Propene	107	1.0	(0.1–2.5)
Surface propene	54	1.3	(0.4–2.5)
Propane	107	0.5	(0.2–0.8)
Surface propane	54	0.4	(0.2–0.8)

associated with the more northern stations of the study area near shore, and were probably related to more direct influences from riverine and estuarine factors. A relatively unique occurrence of higher methane concentrations was routinely observed in the deeper waters of Station 3, Transect IV, during this study. These higher concentrations were attributed to natural gas seepage across the mud-water interface at this southern point in the study area.

Although the highest near-bottom methane concentrations in the southern part of the study area were assumed to be related to natural seepage, other areas of the shelf did exhibit methane maxima. These were typically associated with a bottom nepheloid layer that is often observed, especially during the summer, on the Texas shelf (Figure 12). Although simultaneous measurements of transmissometry and LMWH were only obtained at a few stations, it appears that nepheloid layers are common, especially at Stations 1 and 2 (and bank stations). Most near-bottom methane maxima in these nepheloid layers were accompanied by small increases in ethane levels. It is uncertain whether high LMWH levels in nepheloid layers resulted from resuspension of bottom sediments containing high LMWH levels or from in situ production associated with the layers or both. Higher nutrient and productivity levels associated with these layers may result in high in situ production rates.

Table 4 lists concentrations of ethene, ethane, propene, and propane during 1976 and 1977. The unsaturates dominated over their saturated analogs in most areas of the STOCS, with exceptions gener-

Obser- vations	1977 Mean (nl/l)	(Min.−Max.) (nl/l)
328	239	(41−4000)
54	112	(44−578)
304	4.5	(0.1−21)
54	4.2	(1.9−10)
273	0.7	(0.1−4.6)
53	0.5	(0.1−1.6)
172	1.0	(0.3−2.6)
54	1.2	(0.4−2.6)
170	0.5	(0.2−1.5)
53	0.4	(0.2−1.3)

ally occurring at water depths greater than 100 m. Propene concentrations were almost always four times lower than ethene concentrations. There was good agreement in 1976 and 1977 between average olefin concentrations.

The level of total dissolved and particulate organic matter generally found in the Gulf of Mexico aquatic ecosystem is in the range of 0.1 to 1.0 µg/l. Accurate knowledge at the initiation of this study had not been obtained on ranges of hydrocarbon values in these same waters. Recent data collected from other systems (McAullife 1976; Koons 1977) have indicated that generally the highest concentrations of hydrocarbons are located in the surface microlayer of the water column and that these concentrations decrease rapidly within the first 10 m of water depth.

Concentrations of dissolved and particulate high molecular-weight hydrocarbons were similar in magnitude (Table 5). Particulate hydrocarbon concentrations generally decreased with distance offshore. Higher concentrations of particulate hydrocarbons at inshore stations appeared to result from terrigenous input through direct addition of particulates and increased primary production at shallower sites. Dissolved hydrocarbon concentrations showed less variation. Concentrations showed higher averages for winter and spring than for fall samples. The proportion of dissolved and particulate hydrocarbons varied similarly. The most abundant n-alkanes were in the $C_{27}-C_{33}$ range, with greatest abundances observed for odd carbon numbers.

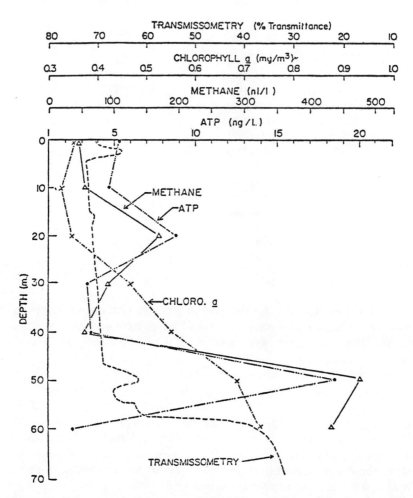

Figure 12. Presence of a bottom nepheloid layer as shown by a depth profile of methane, ATP, chlorophyll *a*, and transmissometry in the STOCS area near Station 2, Transect II, for 15 September 1976.

Table 5. Average Total Hydrocarbons in Seawater by Station (Depth) for All Transects

Station and Year	Dissolved and Particulate ($\mu g/l$)	Total Hydrocarbons Particulate ($\mu g/l$)	Dissolved ($\mu g/l$)
Station 1			
1975	0.43	0.11[a]	0.16[a]
1976	0.34	0.13	0.20
1977	0.46	0.35	0.11
Avg	0.41	0.20	0.16
Station 2			
1975	0.31	0.10[a]	0.11[a]
1976	0.25	0.06	0.19
1977	0.31	0.11	0.20
Avg	0.29	0.09	0.17
Station 3			
1975	0.20	0.05[a]	0.15[a]
1976	0.21	0.05	0.16
1977	0.33	0.12	0.21
Avg	0.25	0.07	0.17

[a] Fall season only.

3
PELAGIC BIOTA

with contributions by P. N. BOOTHE, D. L. KAMYKOWSKI, J. D.
MCEACHRAN, E. T. PARK, P. L. PARKER, L. H. PEQUEGNAT, B. J.
PRESLEY, R. S. SCALAN, P. TURK, J. K. WINTERS, J. H. WORMUTH

Phytoplankton

The hydrographic environment of any marine shelf ecosystem can be very complex because there are so many factors capable of influencing the dynamics of the system. The STOCS area is no exception, with influences such as western land masses and associated riverine input, the deeper offshore Gulf waters, and possibly most important, the Mississippi River to the north. Previous study both of the Texas shelf and of other shelf ecosystems has suggested that the hydrographic features and many of the biotic components, especially pelagic aspects, are strongly correlated. Hydrographic variables, such as temperature, salinity, and currents presented in the preceding chapter, illustrate the annual progression that occurs over the south Texas shelf and help to suggest possible factors that influence the functioning of the ecosystem.

General patterns in phytoplankton biomass on the Texas shelf during the 1975–1977 study period are best illustrated by examining changes at the three stations on Transect II that were sampled monthly 1976–1977. Figures 13 through 15 summarize the temporal and depth patterns in the nanno (A), net (B), and total (C) components of chlorophyll a at three stations of Transect II. Station 1/II (Figure 13) is temporally characterized by a continuous background concentration of nannochlorophyll a; concentration peaks occur in the April and fall (Hurricane Anita) cruises of 1977. Net chlorophyll a is much more variable exhibiting a seasonal peak occurrence between November and May. The seasonality in total chlorophyll a concentration is dominated by the net fraction. Surprisingly, the water column is routinely

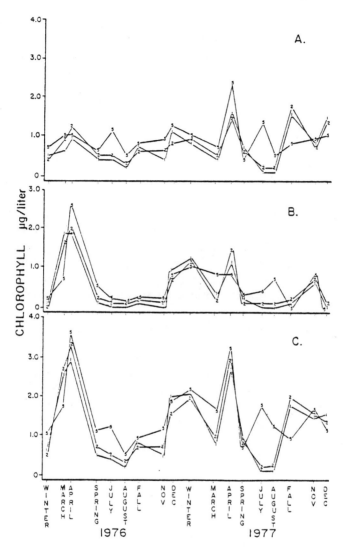

Figure 13. Inner-shelf (Station 1) chlorophyll *a* (µg/l) at the surface (1), half the depth of the photic zone (2), and bottom (5) in the nanno (A), net (B), and total (C) categories plotted against the sampling period.

inverted in chlorophyll *a* concentrations, that is, the maximum concentration occurs in the bottom waters.

Station 2/II (Figure 14) exhibits less variability in the nannofraction than Station 1/II. The concentration of nannochlorophyll *a*, however, generally exceeds that of the net fraction. Two exceptions are at the surface, April 1976, and bottom, July 1977. The total chlorophyll *a* concentration reflects the nanno trend except during the net fraction peaks. The vertical profile of chlorophyll *a* again routinely exhibits an increase with depth.

Station 3/II (Figure 15) exhibits a further decrease in nannochlorophyll *a* variability. An exception occurs at half the depth of the photic zone, winter 1977. The net fraction is extremely low except for the winter of 1977. The total chlorophyll *a* category reflects the even distribution of the other two categories throughout the sampling period except for the combined nanno and net peaks. This unusual concentration of chlorophyll *a* at all depths is related to an upwelling during February 1977. This event may occur every year but may be easily missed in most sampling programs because of its probable short duration.

Analysis of variance (ANOVA) results of the general chlorophyll *a* trends illustrated above showed that the cross-shelf gradient was statistically significant ($P < 0.05$) for all components of chlorophyll at all depths. The bottom samples, however, exhibited a slightly different pattern from the surface or half-depth photic zone collections within these general trends. The latter showed that Station 1 was significantly different from Stations 2 and 3. For bottom water concentrations, on the other hand, the nanno and total chlorophyll categories were similar at Stations 1 and 2, while both these collection sites differed from Station 3.

There was also a north-south chlorophyll gradient observed on the shelf, although it was not as strong as the cross-shelf gradient. The northern part of the STOCS was significantly higher ($P < 0.05$) in chlorophyll *a* both at the surface and at half the depth of the photic zone than at the southern part of the shelf. Measures of chlorophyll in the bottom waters, however, did not show a north-south gradient. These patterns may reflect Mississippi River influences on the shelf, influences that significantly decrease in their impact on the southern collection sites. That the bottom waters do not show the same patterns illustrates lack of mixing on the outer shelf.

Figure 16 summarizes the temporal patterns in the nanno (A), net (B), and total (C) components of carbon-14 uptake at the three stations of Transect II during 1977. Stations 2 and 3 dominate the winter

Pelagic Biota [39]

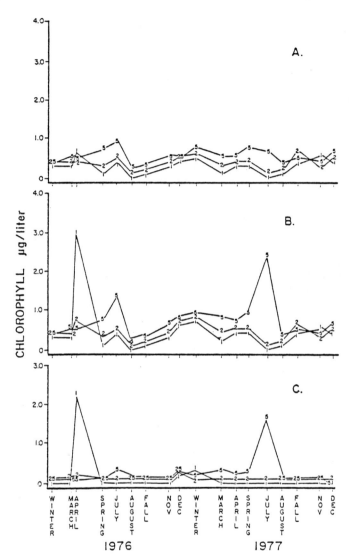

Figure 14. Mid-shelf (Station 2) chlorophyll *a* (μg/l) at the surface (1), half the depth of the photic zone (2), and bottom (5) in the nanno (A), net (B), and total (C) categories plotted against the sampling period.

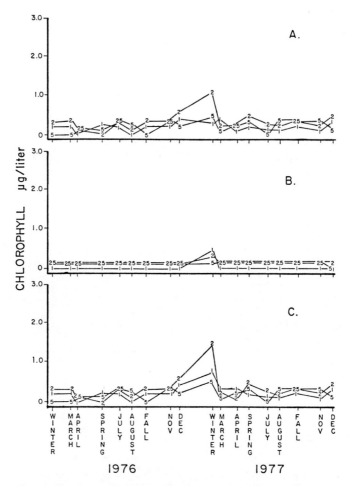

Figure 15. Outer-shelf (Station 3) chlorophyll a (μg/l) at the surface (1), half the depth of the photic zone (2), and bottom (5) in the nanno (A), net (B), and total (C) categories plotted against the sampling period.

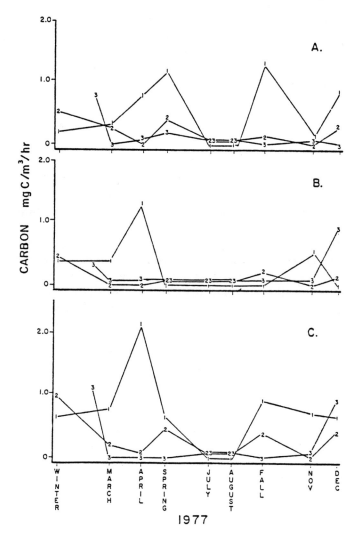

Figure 16. Stations 1, 2, and 3 surface carbon-14 uptake (mgC/m^3/hr) in the nanno (A), net (B), and total (C) categories plotted against the sampling period.

[42] Pelagic Biota

nanno activity; Station 1 is dominant over the majority of the rest of the year. The inshore peaks in nanno activity occur in spring and fall. Station 1 dominates the net activity during the spring and November; Station 3 dominates during December. The peaks in net activity are in April and November. The net peak precedes the nanno peak in the spring bloom and follows it during the fall bloom. The total category presents the composite of the size fractions and provides a picture of classic temperate zone phytoplankton activity.

The chlorophyll concentrations observed during this study, especially in the shallow shelf waters plus the phytoplankton activity represented by Figure 16 suggested that the Texas shelf might be extremely productive in terms of primary producer biomass. Utilizing a technique developed by Ryther and Yentsch (1957), which estimates primary production based upon chlorophyll a measures and light transmittance through the water column, we developed a two-year production curve for Station 1 of Transect II (Figure 17). Primary production for Texas inner-shelf waters is somewhat bimodal annually with peaks in the spring and fall. Annual estimates of production, based upon the chlorophyll a measures converted to carbon equivalents, indicated that these waters produced a mean of approximately

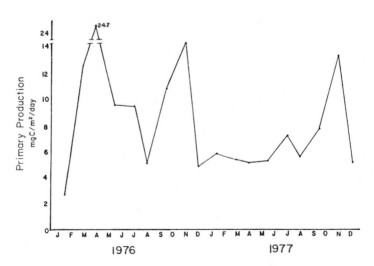

Figure 17. The two-year cycle of primary production (carbon fixation) for Texas inner-shelf waters in 1976 and 1977. Carbon fixation estimated from chlorophyll measures according to technique of Ryther and Yentsch (1957).

103 gC/m²/yr. In contrast, estimates of primary production for coastal waters of other continental shelves that support substantial fisheries, as the STOCS does, such as the North Sea (Steele 1974), indicate an annual production of 70 to 90 gC/m²/yr. It appears that the Texas shelf can be considered a relatively productive ecosystem in terms of plant biomass.

In terms of species composition, the phytoplankton community structure of the STOCS area is complex but relatively consistent in relation to the different water masses that occur on the shelf over the annual cycle. In general, the progression of community structure through the seasons occurs at different rates at different locations on the shelf. The results of cluster analyses have indicated that the temporal changes in community structure of the phytoplankton are related to light intensity, day length, temperature, salinity, stratification, wind, and nutrient sources. The patterns observed over the study period demonstrate the complexity of phytoplankton response to conditions on the Texas shelf.

Species groupings derived from the cluster analyses of phytoplankton are less informative than the station groupings. This condition results for both technical and biological reasons. Technically, the phytoplankton counts are generally limited to the size fraction above 20 μm. Since this fraction is dominant only between December and April, the species groupings represent successions only within this time period. The cruises were not sufficiently frequent to adequately distinguish community changes within this limited period. Information on summer community structure was also limited because the greatest concentration of phytoplankton occurred near the bottom from which species composition samples were not available. The biological reasons are related to the low sampling frequency compared to the rate of change of phytoplankton species composition. The species lists are usually very different from one cruise to the next. Considering these problems, Figure 18 depicts the seasonal patterns of the phytoplankton classes, and Figure 19 depicts the seasonal pattern of selected phytoplankton species or genera from the net phytoplankton observed at the surface along Transect II during 1976 and 1977. The graphs are ordered top to bottom by decreasing numerical abundance.

Diatoms, dinoflagellates, and silicoflagellates are generally most abundant between the November and spring cruises through the winter months (Figure 18). The remaining time interval is represented by a minor dinoflagellate peak, coccolithophorids, and blue-green algae.

Figure 19 demonstrates the seasonal preference of selected taxa.

[44] Pelagic Biota

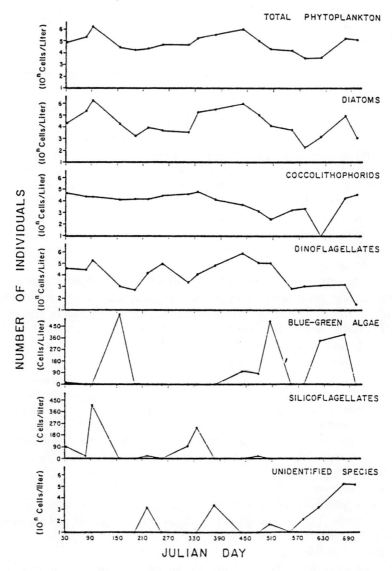

Figure 18. Comparison of seasonal abundances of the different classes of phytoplankton observed along Transect II during 1976 and 1977. Each point represents the sum of the surface abundance at Stations 1, 2, and 3, Transect II, within a sampling period. Number of individuals (cells/l) are plotted against a Julian day calendar. The graphs are ordered top to bottom by decreasing abundance.

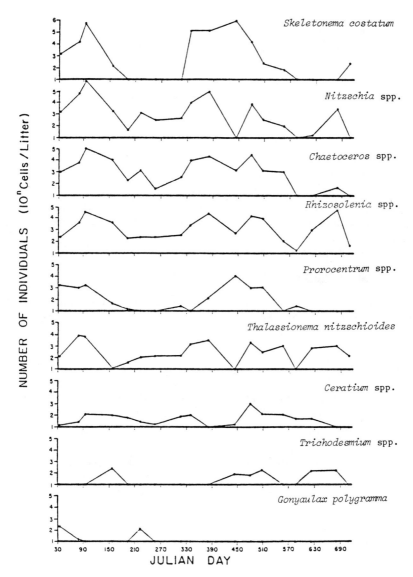

Figure 19. Comparison of seasonal abundances of different species or genera of phytoplankton observed along Transect II during 1976 and 1977. Each point represents the sum of the surface abundances at Stations 1, 2, and 3, Transect II, within a sampling period. Number of individuals (cells/l) are plotted against a Julian day calendar. The graphs are ordered top to bottom by decreasing abundance.

The two years are different in the order of species appearance. In 1976, a relatively clear succession occurs: winter, *Gonyaulax polygramma* and *Prorocentrum micans*; March, *Thalassionema nitzschioides*; April, *Skeletonema costatum*, *Nitzschia* spp., and *Chaetoceros* spp.; spring, *Trichodesmium* spp.; August, *Gonyaulax polygramma* and *Thalassionema nitzschioides*. In 1977, more co-occurrence is evident: pre-March, *Chaetoceros* spp., *Rhizosolenia* spp., *Thalassionema nitzschioides*, *Ceratium* spp., and *Trichodesmium* spp.; November, *Rhizosolenia* spp.

The patterns present a confused picture of the phytoplankton community. These patterns probably result from the complex hydrography in the STOCS area. Better information on the specific dynamics according to shelf-influencing factors may be obtained by eliminating geographic stations and relating species assemblages in similar water masses as was done with the chlorophyll a observations presented above.

Nepheloid Layer

As stated previously, the highest concentrations of chlorophyll a were often observed in the bottom waters of the shelf (Figure 13), especially at the shallow stations on the shelf. The bottom water is also characterized by a pervasive nepheloid layer (Berryhill 1977) at least during part of the annual cycle. Data from several cruises in 1978 to examine nepheloid layer dynamics not only demonstrated a prevalent nepheloid layer (Figure 20), but also illustrated the presence of peak chlorophyll levels in the bottom waters, as well as peaks in nitrogen represented by ammonia. These peaks of primary producer biomass as well as greater than 1% light transmissions at these depths suggested the possibility of photosynthesis taking place. Carbon-14 experiments confirmed this (Kamykowski and Batterton 1979). In addition to the primary producer biomass in bottom waters, there appeared to be a considerable amount of nutrient regeneration as illustrated by the ammonia concentrations.

The basic conclusions drawn from the study of the nepheloid layer during four diel sampling cruises in 1978 were

1. The nepheloid layer was present throughout the 24-hr sampling period and fluctuated in thickness and density within this period;
2. Phytoplankton are concentrated in the nepheloid layer during the summer months in the STOCS area. Active carbon fixation can occur since 10% surface radiation may often reach the sediment interface in the zone within 50 km of shore;

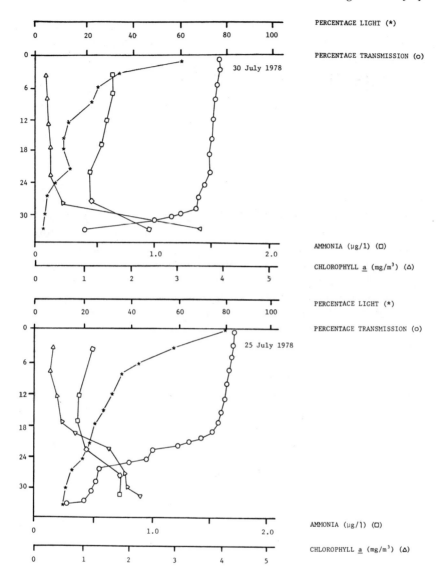

Figure 20. Depth profiles of percentage light and transmissometry and chlorophyll *a* and ammonia nitrogen concentrations for two cruises off the Texas coast (33-m water depth) during 1978 near Station 4, Transect II.

[48] Pelagic Biota

3. Since nutrients are probably supplied to the layer at least partly from benthic diffusion, the phytoplankton dynamics of the layer may be affected by perturbations of the benthos caused by oil-related activities; and
4. The overall impact of this effect depends on organism sensitivity, the area perturbed, exchange intensity, and the trophic significance.

Neuston

The neuston is defined as plants and animals that live on or just beneath the surface film of marine waters. Sargassum mats are usually associated with this surface component of the Texas shelf waters as well as some freshwater plants, such as water hyacinth that enters the system through riverine inflow, especially farther south near the United States–Mexico border. Very diverse communities of fauna are normally associated with these floating mats. The focus of the STOCS study, however, was on the more free-living fauna that inhabits this surface zone. It is believed that many potential pollutants that enter the marine system do so through the surface waters (e.g., petroleum hydrocarbons), and any biological impact from these pollutants may first manifest itself in changes of the neuston.

Although the neuston defies a strict biological definition in terms of species, there are certain taxonomic groups that are commonly found in the upper 15 to 20 cm of the water column during significant portions of each day. There is considerable variability not only in the abundance of neuston, either as total numbers of organisms or dry weight, as well as taxonomic composition. This is due, in part, to diel vertical migration but also to various types of environmental heterogeneity. Day-night sampling helps to minimize diel variation, but environmental heterogeneity is not generally monitored.

The number of organisms and densities of these organisms collected from all stations for each season varied widely both temporally and within taxa (Table 6). Neuston biomass was also highly variable during the study interval. Most taxa showed distinct seasonal cycles with peaks occurring during the spring and summer sampling periods. In contrast, a few late fall-winter species were found. These general trends showed good year-to-year reproducibility. Onshore-offshore variation was observed in the distribution of some taxa, particularly the larval decapod crustaceans.

The neuston decapod fauna was studied in great detail during 1976 and 1977. A total of 104 decapod taxa were identified, consisting of 88 larval taxa and 16 nonlarval taxa. Decapod larvae accounted for 53% of the mean concentration of total decapods and 6% of the total neuston.

Decapod larval species diversity was greater in spring and fall than in winter and greater at stations near shore than at offshore stations. The decrease in larval diversity with distance from shore could be expected since decapod larval input into the surface waters is greatest over inshore areas where benthic decapod adult populations are more diverse and estuarine input is a direct influence on the system. Nearly all of the dominant decapod larvae reached greatest concentrations during the spring.

A large number of fish taxa occur in the neuston off south Texas for at least part of their life span; most have distinct seasonal, diel, and horizontal distribution cycles. The neuston fauna consisted of a cold water component, present either from fall through winter or from winter through early spring; a warm water component, present either from late spring through summer or entirely during the summer; and a ubiquitous component, present in high abundance most of the year. Within each of the seasonal components, taxa were generally distributed either inshore or offshore and were present in the neuston zone either nocturnally or diurnally.

Diversity of fish taxa, when computed for each of the sampling years, was relatively high ($H' = 2.72$ for 1976; $H' = 2.58$ for 1977). In 1976, the most abundant taxa (those that individually represented 2.5% or more of the total) were *Antennarius* sp. (22.6%), *Harengula jaguana* (11.9%), *Mugil cephalus* (8.6%), Mullidae (8.1%), *Opisthonema oglinum* (4.4%), *Cynoscion* sp. (4.2%), Gerreidae (3.8%), *Engraulis eurystole* (3.0%), *Micropogon undulatus* (2.9%), and *Citharichthys spilopterus* (2.5%). With the exception of *Antennarius* sp., which was captured at only three stations, these taxa were widely distributed over the survey area during at least one of the sampling seasons (winter, spring, fall).

In 1977 the most abundant taxa, representing 2.5% or more of the total, were Mullidae (18.1%), *Etrumeus teres* (12.9%), *Harengula jaguana* (8.5%), Gerreidae (4.7%), *Trachurus lathami* (4.6%), *Rachycentron canadum* (4.0%), *Mugil curema* (3.3%), *Prionotus* spp. (3.2%), *Mugil cephalus* (2.6%), and *Menticirrhus* spp. (2.6%). All of these taxa, with the exception of *Rachycentron canadum*, were widely distributed over the survey area during at least one of the sampling seasons (winter, spring, fall). The Gerreidae, *Prionotus* spp., and *Menticirrhus* spp. were each represented by several species in the survey area.

Analysis of variance of tar concentrations collected in neuston tows

[50]   Pelagic Biota

Table 6. Mean Abundances of Selected Neuston Taxa by Cruise and Time of Day[a]

Cruise		N^b	Hyperidae			Lucifer faxoni			Brachyuran megalopa			Brachyuran zoea		
			u	$\bar{X}$	l	u	$\bar{X}$	l	u	$\bar{X}$	l	u	$\bar{X}$	l
Winter	Day	12	—	413	—	—	3,988	—	—	447	—	255	152	48
	Twilight	—	—	—	—	—	—	—	—	—	—	—	—	—
	Night	12	31,905	17,176	2,447	4,381	2,874	1,367	—	11,897	—	6,650	3,962	1,273
March	Day	3	—	84	—	—	125	—	—	165	—	—	551	—
	Twilight	—	—	—	—	—	—	—	—	—	—	—	—	—
	Night	3	—	31,349	—	—	5,844	—	—	13,620	—	—	11,593	—
April	Day	3	—	343	—	—	230	—	—	23	—	—	327	—
	Twilight	—	—	—	—	—	—	—	—	—	—	—	—	—
	Night	3	—	14,443	—	—	9,684	—	—	1,584	—	—	12,903	—
Spring	Day	6	—	805	—	—	2,609	—	—	696	—	—	1,494	—
	Twilight	7	—	36,624	—	115,152	60,424	5,695	—	430	—	—	6,159	—
	Night	11	151,055	85,268	19,480	19,761	13,132	6,502	22,427	12,698	2,968	18,182	12,021	5,859
July	Day	1	—	99	—	—	35	—	—	105	—	—	738	—
	Twilight	3	—	99	—	—	9,285	—	—	2,768	—	—	426	—
	Night	2	—	4,554	—	—	3,638	—	—	8,129	—	—	2,737	—
August	Day	2	—	75	—	—	101,362	—	—	1,659	—	—	3,127	—
	Twilight	1	—	89	—	—	757	—	—	—	—	—	178	—
	Night	3	—	29,355	—	—	6,426	—	—	11,105	—	—	9,214	—
Fall	Day	11	—	2,653	—	5,634	3,130	625	—	325	—	—	—	—
	Twilight	1	—	636	—	—	29,771	—	—	14,808	—	—	—	—
	Night	12	53,563	28,495	3,877	5,528	3,502	1,475	10,800	6,169	1,538	4,931	2,775	619
November	Day	2	—	2,411	—	—	15,637	—	—	1,269	—	—	379	—
	Twilight	1	—	5,451	—	—	1,777	—	—	7,229	—	—	355	—
	Night	3	—	6,184	—	—	1,874	—	—	4,001	—	—	701	—
December	Day	2	—	817	—	—	22,024	—	—	815	—	—	1,838	—
	Twilight	1	—	41	—	—	57	—	—	—	—	—	131	—
	Night	3	—	1,898	—	—	1,385	—	—	4,452	—	—	2,824	—

Pelagic Biota [51]

Cruise		N[b]	Calanopia americana females			Centropages velificatus females			Nannocalanus minor			Temora stylifera		
			u	X̄	l	u	X̄	l	u	X̄	l	u	X̄	l
Winter	Day	12	—	58	—	—	480	—	—	616	—	—	1,592	—
	Twilight	—	—	—	—	—	—	—	—	—	—	—	—	—
	Night	12	2,068	1,331	594	2,513	1,537	560	14,691	8,520	2,349	3,131	1,904	676
March	Day	3	—	0	—	—	478	—	—	358	—	—	1,638	—
	Twilight	—	—	—	—	—	—	—	—	—	—	—	—	—
	Night	3	—	1,661	—	—	2,155	—	—	21,792	—	6,726	6,542	6,357
April	Day	3	—	27	—	—	108	—	—	40	—	—	218	—
	Twilight	—	—	—	—	—	—	—	—	—	—	—	—	—
	Night	3	—	3,317	—	10,563	5,808	1,052	—	16,428	—	—	17,273	—
Spring	Day	6	—	11	—	—	29,311	—	—	1,071	—	6,817	3,836	854
	Twilight	7	—	4,165	—	—	59,366	—	—	—	—	179,432	97,112	14,792
	Night	11	—	930	—	72,905	41,955	11,004	4,082	2,176	269	77,722	47,760	17,797
July	Day	1	—	—	—	—	1,265	—	—	990	—	—	4,219	—
	Twilight	3	—	—	—	—	114	—	—	—	—	—	518	—
	Night	2	—	123	—	—	1,294	—	—	1,265	—	—	6,027	—
August	Day	2	—	—	—	—	3,319	—	—	133	—	—	671	—
	Twilight	1	—	445	—	—	757	—	—	240	—	—	3,208	—
	Night	3	—	5,215	—	—	3,885	—	—	—	—	—	1,579	—
Fall	Day	11	—	24	—	—	1,491	—	—	9	—	1,408	780	151
	Twilight	1	—	159	—	—	955	—	—	—	—	—	159	—
	Night	12	—	4,529	—	11,929	6,517	1,105	—	526	—	—	1,311	—
November	Day	2	—	677	—	—	5,068	—	—	—	—	—	507	—
	Twilight	1	—	829	—	—	177	—	—	2,666	—	—	118	—
	Night	3	—	4,507	—	—	8,144	—	—	489	—	—	365	—
December	Day	2	—	3,071	—	—	2,405	—	—	—	—	—	1,514	—
	Twilight	1	—	8	—	—	1,748	—	—	—	—	—	41	—
	Night	3	—	22,433	—	—	11,549	—	—	2,554	—	—	3,360	—

Table 6—Continued

Cruise		N^b	Anomalocera ornata immatures			Labidocera immatures			Pontellopsis villosa males			Chaetognaths		
			u	$\bar{X}$	l	u	$\bar{X}$	l	u	$\bar{X}$	l	u	$\bar{X}$	l
Winter	Day	12	20,370	11,782	3,193	—	—	—	—	177	—	2,998	1,804	609
	Twilight	—	—	—	—	—	—	—	—	—	—	—	—	—
	Night	12	9,567	4,793	19	—	347	—	—	40	—	11,878	8,776	5,674
March	Day	3	—	8,264	—	—	1,207	—	—	16	—	—	3,130	—
	Twilight	—	—	—	—	—	—	—	—	—	—	—	—	—
	Night	3	—	13,792	—	—	3,169	—	—	110	—	—	21,746	—
April	Day	3	—	966	—	—	13,743	—	—	—	—	—	309	—
	Twilight	—	—	—	—	—	—	—	—	—	—	—	—	—
	Night	3	—	9,233	—	—	100,612	—	—	145	—	10,388	7,731	5,073
Spring	Day	6	—	—	—	—	184,046	—	3,849	2,367	884	7,431	3,821	210
	Twilight	7	—	—	—	—	26,664	—	—	61,450	—	18,362	12,468	6,573
	Night	11	—	—	—	—	5,360	—	—	989	—	34,078	22,998	11,917
July	Day	1	—	—	—	—	984	—	—	35	—	—	1,336	—
	Twilight	3	—	—	—	—	8,089	—	—	516	—	—	2,306	—
	Night	2	—	—	—	—	7,006	—	—	24	—	—	4,539	—
August	Day	2	—	—	—	—	113	—	—	2,428	—	—	720	—
	Twilight	1	—	—	—	—	44	—	—	222	—	—	2,718	—
	Night	3	—	—	—	—	232	—	—	3,714	—	—	4,575	—
Fall	Day	11	—	—	—	995	599	202	2,861	2,184	1,506	—	2,111	—
	Twilight	1	—	—	—	—	159	—	—	636	—	—	477	—
	Night	12	—	—	—	2,372	1,393	414	—	2,036	—	13,456	8,733	4,009
November	Day	2	—	20,098	—	—	590	—	—	1,943	—	—	5,443	—
	Twilight	1	—	—	—	—	237	—	—	—	—	—	1,777	—
	Night	3	—	2,068	—	—	189	—	—	474	—	—	5,412	—
December	Day	2	—	—	—	—	—	—	—	2,968	—	—	5,440	—
	Twilight	1	—	—	—	—	41	—	—	90	—	—	362	—
	Night	3	—	—	—	—	517	—	—	297	—	—	11,350	—

Cruise		N[b]	larvae			eggs		
			u	X̄	l	u	X̄	l
Winter	Day	12	809	427	44	—	14,767	—
	Twilight	—	—	—	—	—	—	—
	Night	12	852	577	302	13,663	—	472
March	Day	3	—	430	—	—	9,118	—
	Twilight	—	—	—	—	—	—	—
	Night	3	—	3,068	—	—	6,951	—
April	Day	3	—	233	—	—	2,171	—
	Twilight	—	—	—	—	—	—	—
	Night	3	1,061	722	382	—	5,307	—
Spring	Day	6	—	341	—	—	4,114	—
	Twilight	7	—	543	—	—	7,182	—
	Night	11	2,597	1,601	604	6,348	3,507	—
July	Day	1	—	8	—	—	1,371	—
	Twilight	3	—	205	—	—	1,497	—
	Night	2	—	2,555	—	—	6,247	—
August	Day	2	—	79	—	—	1,706	—
	Twilight	1	—	133	—	—	757	—
	Night	3	—	3,584	—	—	5,299	—
Fall	Day	11	136	72	7	—	741	—
	Twilight	1	—	62	—	—	159	—
	Night	12	1,656	961	265	965	534	102
November	Day	2	—	357	—	—	866	—
	Twilight	1	—	125	—	—	59	—
	Night	3	—	424	—	—	429	—
December	Day	2	—	3,859	—	—	477	—
	Twilight	1	—	113	—	—	—	—
	Night	3	—	1,405	—	—	700	—

[a] Mean, X̄. Numbers are given in number/10³m³. Upper (u) and lower (l) 95% confidence limits are given only when they do not cross zero.
[b] N, number of observations.

[54]   Pelagic Biota

showed significant differences according to season and transect but not station or time. For example, tar concentrations were highest during March and September–November. In part, this was due to single high values that shifted the means upward. Figure 21 shows variations in tarball concentrations with transect in the STOCS area. Highest averages were observed on Transect I and II with dramatic decreases on Transects III and IV. It should be noted that Transect II was sampled more consistently (monthly) during the study period, which may account for the higher mean concentrations. The trends suggest, however, that presently the surface concentrations of hydrocarbons may be related to ship traffic in the Aransas Pass Inlet and other points in the northern Gulf and extensive petroleum activities in the waters off Louisiana, north of the STOCS study area.

The only significant correlation for neuston biomass observed during this study was to the amount of tar obtained in the same samples. Two theories may possibly explain this phenomenon. One concerns surface circulation that creates convection cells, that is, Langmuir cells (Pollard 1977). If tar is considered to be a passively floating object, then it would be expected to be concentrated by this type of circulation into windrows. The positive correlation then may suggest

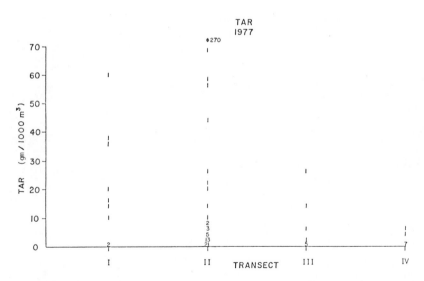

Figure 21. The variation in tar concentration by transect. The numbers represent the number of coincident points.

that neuston, in general, is also concentrated in windrows to some extent.

An alternative explanation may be related to the nutrition of neuston. It has often been observed in Gulf waters that small fish and larger crustaceans (i.e., crabs) are frequently associated with small floating pancakes of mousse and tarballs. These reports have included observations of the fauna feeding off the tar and mousse. There is a very good likelihood that well-weathered petroleum products floating in the Gulf surface waters develop growths of epiphytes and other colonial forms normally associated with hard surfaces. These growths on the floating tar could possibly be providing a food source to many surface-oriented species, including neuston.

Neuston cluster grouping results showed patterns consistent with water temperature trends. Thus it could be concluded that the dynamics of the neuston may be controlled by temperature, which is the influential factor related to spawning of many of the temporary populations found in the neuston.

In general, the faunal composition of the neuston differed from that of the water column below the neuston zone. Finucane (1976, 1977) reported on the ichthyoplankton captured in the water column from the same stations sampled in this study and found a number of differences between the species composition of the two. Fishes of the neuston zone can be classified as facultative neustonic taxa or euneustonic taxa. Those of the former category are found in the neuston during the larval stage, while those of the latter category spend their entire lives in the neuston. Facultative forms dominate the neuston, and a majority spend their juvenile and adult stages in the estuaries and inshore waters of the northwestern Gulf of Mexico.

Diel variability played a large role in influencing the dynamics of the STOCS neuston. In addition, distance from shore played a role for many taxa, particularly the decapods, probably due in large part to estuarine influences and the benthic distribution patterns of the adult species.

Zooplankton

Zooplankton biomass, total density, and female copepod density decreased with distance offshore. When only the means were considered (Figures 22 and 23), biomass weights not only decreased seaward but varied most consistently between seasons at the shallow stations. Mean total zooplankton densities varied similarly in a seaward direction, but seasonal fluctuations by depth were poorly patterned. Mean

[56] Pelagic Biota

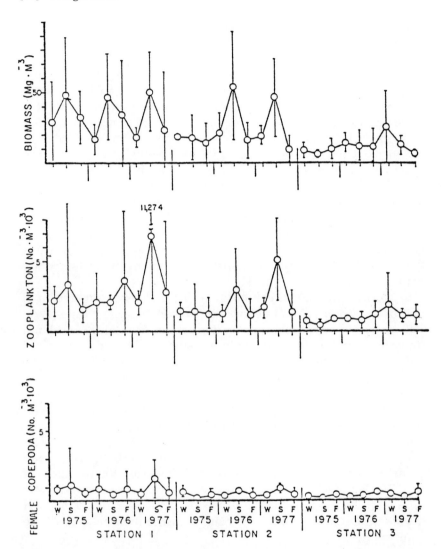

Figure 22. The mean (circles) and 95% confidence interval (bars) for representative zooplankton abundance variables (average of four transects for each station and season).

Pelagic Biota [57]

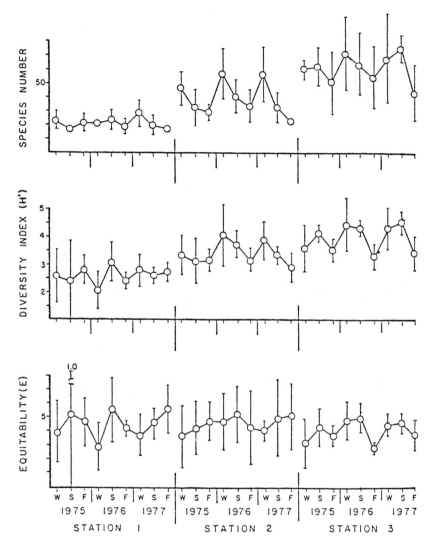

Figure 23. The mean (circles) and 95% confidence interval (bars) for representative zooplankton community structure variables (average of four transects for each station and season).

densities of female copepods changed with the total zooplankton, decreasing with depth. Species diversities followed the same patterns of change with depth and season that were observed in the number of species. The evenness of population distributions as measured by equitability, on the other hand, showed almost no pattern of change related to depth or season. The relatively low mean values obtained for equitability at each depth indicated that a few species accounted for most of the zooplankton density across the shelf.

The dominant female copepod species observed during this study are listed in Figure 24 along with their frequency of occurrence at three stations along Transect II. Copepod species formed five species groups that were generally related to water depth (station location) on the transect. The first group was generally considered ubiquitous in nature. Group 2 was confined to the shallow waters on the shelf, Group 3 to middepths, and Groups 4 and 5 to the deep waters. In addition to the general trends observed in Figure 24, the dominant species at both the shallow and mid-depth stations on the Texas shelf usually differed during the semiannual periods December–June and July–November.

The results of measuring representative zooplankton variables such as biomass, total zooplankton density, and female copepod density revealed considerable variation in zooplankton distribution along bottom-depth contours from transect to transect during each of the nine seasonal cruises. The variability suggested the occurrence of pulsing input to the systems that encouraged zooplankton production but that was so limited that the entire length of the study area was not uniformly affected. For example, the calanoid species *Acartia tonsa*, *Paracalanus indicus*, *Paracalanus quasimodo*, and *Clausocalanus furcatus* were often found in dense patches on one or two transects in the spring. Cladocerans, *Penilia* spp., appeared in a highly regionalized dense patch at Station 1, Transect II, in the spring of 1977 and in August 1976. Ostracods, primarily *Euconchoecia chierchiae*, were found in dense patches at various stations throughout the study.

It is possible that the patchy distribution of the zooplankton in the study area was related to pulses of low salinity input from bay systems. Evidence for estuarine influence in the STOCS area may be found in the composition of copepod species. *Acartia tonsa* is a calanoid copepod that is almost always reported among the most abundant copepod species inhabiting bays and estuaries in the Gulf of Mexico and along the Atlantic coast from Florida to Cape Hatteras (Breuer 1962; Cuzon du Rest 1963; Bowman 1971). In the STOCS area *Acartia tonsa* appeared in large numbers in 1975 at the stations near

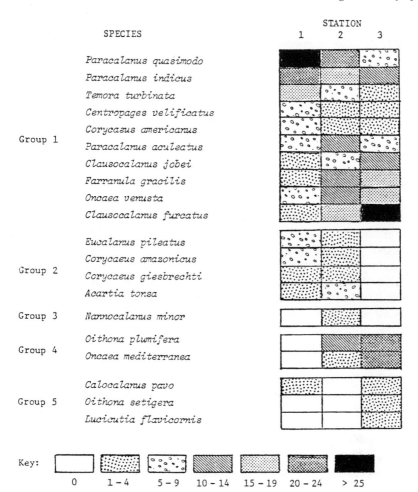

Figure 24. The frequency of occurrence of female copepod species used in cluster analysis for each bottom depth (Stations 1–3 are in order of increasing depths). Maximum number of occurrences is 36.

shore and mid depth on Transects I and II in the spring. In all three years, *Acartia tonsa* was most abundant in the spring when salinities were low, and the largest abundances in other typically estuarine copepod species (*Centropages hamatus, Labidocera aestiva, Oithona nana,* and *Paracalanus crassirostris*) were also most abundant.

Multiple regression analysis identified possible relationships between zooplankton densities and physical, nutrient, and phytoplankton variables. A number of expected, or at least plausible, relationships were indicated; however, occasional relationships between trophically separated entities (i.e., phosphates and copepods) suggested that some of the relationships may be deceptive. At the shallow stations, changes in ichthyoplankton more frequently were related to the variation in zooplankton variables than any of the other independent (physical or phytoplankton) variables examined. This may indicate that ichthyoplankton populations were possibly responding to changes in zooplankton density. It is generally accepted that some species of planktonic fish take advantage of the zooplankton as a food source (e.g., Peters and Kjelson 1975).

Salinity, although related to several zooplankton community characteristics at the shallow stations, was more often highly correlated to these at the mid-depth stations. The implied relationships between zooplankton and salinity at the mid-depth stations may indirectly reflect a response of the zooplankton to changes in primary production that have been shown to be commonly associated with salinity changes in neritic waters.

The number of associations between zooplankton and phytoplankton variables increased seaward. At the deep stations, phytoplankton density generally accounted for the largest percentage of explained variation in zooplankton variables. The implied direct relationship of zooplankton to phytoplankton at the deep stations reflects a close dependence of zooplankton on phytoplankton, a relationship generally reported for oceanic, subtropical, or tropical marine waters (Menzel and Ryther 1961; Sander and Moore 1978). The results of this study suggested that offshore zooplankton populations may be controlled by food availability, while zooplankton populations near shore may be controlled by predation.

Results from studies of microzooplankton on the Texas shelf indicated that protozoa reached a maximum in abundance in early spring (March–April). A second protozoan abundance peak was noted in September 1977, but this peak was thought to be atypical and a result of Hurricane Anita, which passed through the area at that time. Oligotrichida was the dominant protozoan group on the STOCS, both spatially and temporally. The other protozoan groups tended to be more restricted both in space and time. Species diversity was high during most of the year and varied erratically. Protozoan biomass ranged from 1% to 348% of the macrozooplankton biomass, indicating that protozoa are a significant component of the zooplankton community.

Zooplankton Body Burdens

HYDROCARBONS Approximately 50% of the zooplankton hydrocarbon samples taken in 1977 showed the possible presence of petroleum-like organic matter. This percentage was slightly higher than that observed for samples in 1976 (30%) and considerably higher than that observed in 1975 (7%). This apparent increase may be a reflection of the increased importing of crude oils during this time.

The criteria for presence of petroleum-like organic matter are smooth distribution in the region of n-alkanes of molecular size greater than C_{21} and odd-even preference (OEP) values close to unity. In the case of samples analyzed by gas chromatography and mass spectrometry (GC/MS) techniques, the presence of aromatic compounds is usually indicative of petroleum-like material.

Seven zooplankton samples were investigated by GC/MS analyses in 1977. One sample from 3/IV in spring contained polynuclear aromatic hydrocarbons (PAH) in quantities such that they were readily identified, even though the quantities were inadequate for the components to be observable in the gas chromatographic analysis. Four samples (1/II, spring; 2/III, spring; 3/II, spring; and 1/IV, spring), showed possible trace quantities of PAH by GC/MS analysis, though quantities were inadequate to permit certain identification. Two samples (2/IV, winter and 1/III, fall) showed no indication of the presence of PAH. All seven of these samples met the n-alkanes distribution criterion as possibly having petroleum-like organic matter present.

Although relatively high OEP values were observed for particulate hydrocarbons in the water column, these values were considerably lower than values found for zooplankton in this study. Zooplankton average OEP values for nine seasonal sampling periods ranged from 2.0 to 15.4 with an average of 5.8. The comparatively low values of OEP for particulate hydrocarbons (0.9–1.6) suggested that the majority of these hydrocarbons were not synthesized by zooplankton or higher plants. The higher zooplankton values observed could have reflected the bioaccumulation and concentrating tendencies of zooplankton for pelagic particulate matter in general during their feeding activities. It is well documented (Conover 1971) that zooplankton will ingest microtarballs and other petroleum forms from the water column and pass them through their systems without digesting them. This is also a mechanism for input of petroleum hydrocarbons to the benthos via zooplankton fecal pellets.

The usual range of total hydrocarbon concentration in zooplankton was 50 to 500 µg/g. Total hydrocarbons did not show either temporal or spatial variations that were significant. None of the isoprenoid variables were shown to vary in a statistically significant manner.

Seasonal averages of zooplankton hydrocarbon concentrations in the region of $C_{25}-C_{32}$ (which represent the sum of highest carbon ranges of high molecular-weight hydrocarbons) as well as the seasonal average OEP values are plotted in Figure 25A. Since both variables were estimates of the quantity of petroleum hydrocarbons, they were considered as a single data set, and from the set a curve was constructed. The large standard deviation associated with both parameters results from the "patchiness" of zooplankton and petroleum hydrocarbons likely present as microtarballs. In view of the difficulty in obtaining representative samples composed of two components each having its own unequal distribution, the fit of points to the curve is rather good. The data (Figure 25A) suggest a significant increase in the contribution of petroleum hydrocarbons to zooplankton samples during the three-year study period.

Exploration and drilling activities in the study area probably were not a major source of petroleum residues found in the zooplankton samples. A much more likely source would be the oil tankers that delivered increased quantities of crude oil to Texas ports during the study period. The quantity of crude oil imported to the Port of Corpus Christi and Harbor Island from 1975 through 1977 has been plotted in Figure 25B. The curve generated by the data of Figure 25A has been included in Figure 25B for comparison, and the two show a good correlation suggesting potential causes of hydrocarbon increases.

TRACE METALS   Table 7 summarizes the three years of zooplankton trace element data by station and transect where samples were collected. The only truly meaningful spatial effect observed in zooplankton was an increase in cadmium (Cd) concentrations offshore. The reason for this trend is not clear. The trend does not suggest any significant anthropogenic input of Cd to the environment near shore. Secchi depth, however, was strongly correlated with Cd levels ($r^2 = 0.30$). This variable is a measure of turbidity and suggested that zooplankton Cd levels were influenced in some way by the amount of suspended particulate matter in the water column.

Table 8 summarizes the average seasonal concentrations of trace metals in zooplankton observed during this three-year study. Aluminum (Al), iron (Fe), and nickel (Ni) exhibited significant seasonal trends. Elevated levels of Al and Fe in zooplankton samples are generally interpreted as incorporation of clay particles by the zooplankters (Martin and Knauer 1973). Considerable evidence suggested that

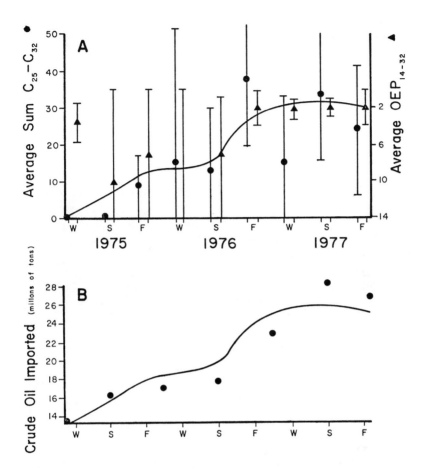

Figure 25. Petroleum hydrocarbons. *A*, average sum of the highest carbon ranges of HMW hydrocarbons (C_{25}–C_{32}) and average OEP of zooplankton samples 1975–1977; *B*, crude oil imported to the Port of Corpus Christi and Harbor Island 1975–1977. Curve from *A* has been included in *B* for comparison.

Table 7. Average Concentrations of Trace Elements in Zooplankton[a]

Transect	Station	No. of Samples	Cd	Cr	Cu	Fe	Ni
			(95% confidence interval observed around mean)				
I	1	18	1.4 (0.65–3.0)	6.0 (0.10–22)	14 (5.0–23)	4,500 (400–13,000)	8.5 (0.60–20)
	2	20	3.0 (1.6–6.0)	4.5 (0.35–14)	21 (6.0–90)	1,900 (100–5,000)	6.0 (2.0–11)
	3	12	5.0 (3.0–7.0)	2.5 (0.40–6.0)	24 (9.5–70)	1,200 (130–3,900)	8.0 (3.0–15)
II	1	16	2.4 (0.95–5.5)	4.0 (0.70–14)	20 (2.5–75)	3,000 (35–13,000)	5.0 (2.2–16)
	2	20	3.5 (1.8–5.5)	3.5 (0.50–9.0)	190 (5–2,500)	2,100 (20–8,500)	7.0 (1.9–30)
	3	12	5.0 (3.5–7.0)	2.5 (0.10–7.5)	21 (7.0–90)	1,600 (40–8,000)	6.5 (2.0–18)
III	1	18	2.0 (0.65–4.0)	4.5 (0.60–13)	16 (5.5–60)	3,000 (240–17,000)	5.5 (0.95–30)
	2	20	3.5 (1.5–5.5)	4.5 (0.30–10)	14 (8.0–20)	3,000 (550–6,600)	7.0 (3.0–18)
	3	12	4.5 (1.8–6.0)	3.5 (0.75–8.0)	13 (6.0–30)	2,500 (350–11,000)	7.0 (2.0–17)
IV	1	16	3.0 (0.80–4.5)	3.0 (0.45–8.5)	13 (6.0–35)	3,000 (200–12,000)	6.0 (2.0–16)
	2	18	3.0 (0.60–4.5)	5.5 (0.11–16)	50 (8.0–300)	5,000 (24–15,000)	10 (2.0–40)
	3	12	4.0 (2.5–6.0)	4.0 (0.10–11)	17 (7.5–55)	550 (70–1,600)	5.5 (3.0–8.)
	Transect[b]		NS	NS	NS	NS	NS
	Station[b]		0.001	NS	NS	0.005	NS

[a] Concentration in ppm dry weight.
[b] ANOVA results: metals for which the main effect indicated was significant at level shown. NS, not significant ($P > 0.05$).

Pelagic Biota [65]

Pb	V	Zn	Al	Ca
(95% confidence interval observed around mean)				
22	21	120	7,000	35,000
(1.8–160)	(4.0–45)	(9.0–500)	(1900–19,000)	(14,000–40,000)
13	9.5	125	2,500	30,000
(1.4–75)	(1.2–20)	(6.0–210)	(140–6,500)	(4,500–60,000)
7.0	7.0	130	2,200	35,000
(1.3–13)	(1.4–25)	(95–160)	(100–10,000)	(22,000–65,000)
11	16	130	5,500	30,000
(1.3–70)	(0.4–70)	(25–250)	(12–25,000)	(16,000–45,000)
11	16	180	4,000	65,000
(1.0–70)	(2.2–65)	(22–500)	(75–14,000)	(8,000–140,000)
12	14	110	2,500	30,000
(0.6–65)	(2.0–25)	(40–190)	(95–12,000)	(16,000–50,000)
15	25	130	7,000	60,000
(0.60–45)	(4.0–60)	(30–270)	(850–30,000)	(18,000–100,000)
10	15	140	6,000	50,000
(0.80–40)	(3.0–70)	(35–500)	(1300–30,000)	(25,000–80,000)
8.5	10	220	4,500	30,000
(0.80–30)	(3.5–35)	(75–1,300)	(300–17,000)	(25,000–35,000)
7.0	13	170	4,500	25,000
(0.80–40)	(4.5–50)	(75–500)	(550–20,000)	(9,500–35,000)
23	24	350	9,000	40,000
(0.55–140)	(2.3–85)	(9.0–2,000)	(80–25,000)	(10,000–70,000)
12	13	180	1,500	4,000
(0.60–40)	(5.0–25)	(80–1,000)	(200–3,000)	(35,000–50,000)
NS	NS	NS	NS	NS
NS	NS	NS	0.008	NS

Table 8. Average Seasonal Concentrations of Trace Elements in Zooplankton[a]

Season[b]	No. of Samples	Cd	Cr	Cu	Fe	Ni	Pb	V	Zn	Al	Ca
		(95% confidence interval observed around mean)									
Winter	56	3.0 (0.85–5.0)	4.0 (0.60–8.5)	15 (4.5–45)	2,300 (120–6,000)	5.5 (2.0–9.5)	15 (1.5–45)	13 (4.0–40)	160 (25–500)	4,500 (250–13,000)	30,000 (9,500–50,000)
Spring	70	3.5 (1.1–6.0)	3.5 (0.10–10)	16 (6.0–38)	950 (23–3,500)	6.0 (1.9–18)	7.5 (0.60–45)	13 (1.3–65)	130 (40–200)	1,300 (75–5,000)	45,000 (16,000–90,000)
Fall	68	3.0 (0.65–6.0)	5.5 (0.15–14)	70 (5.5–210)	5,500 (30–15,000)	9.5 (1.6–25)	14 (0.65–80)	25 (4.0–45)	220 (9.0–1,000)	11,000 (100–25,000)	50,000 (16,000–95,000)
Season[c]		0.022	NS	NS	0.001*	0.001*	NS	NS	NS	0.001*	0.005

[a] Concentration in ppm dry weight.
[b] Seasons: Winter, Jan.–Feb.; Spring, May–June; Fall, Sept.–Oct.
[c] ANOVA results: metals for which season main effect was significant at the level shown. NS means not significant ($P > 0.05$). Asterisk (*) indicates season main effect was significant in both two-way ANOVA tests made that included that effect.

this process is responsible for the seasonal trends observed here. The concentrations of Al and Fe in suspended matter from the Gulf of Mexico were approximately 9% and 5%, respectively (Trefry and Presley 1976a). The Fe:Al ratio in such particulates was 0.56. Aluminum and Fe levels in zooplankton samples from this study were strongly correlated ($r^2 = 0.81$) and the average Fe:Al ratio was 0.52. In addition, the trend in zooplankton Al and Fe concentrations (Table 8) corresponded well to the observed seasonal fluctuations in suspended matter concentrations in STOCS surface waters. Suspended particulate concentrations are generally highest in the fall and lowest in the spring (Berryhill 1978). Also, Al and Fe levels in zooplankton decreased as distance increased from shore. These geographical trends for zooplankton were significant in only one of the two ANOVA tests conducted involving station effect. As a result they could not be considered completely clear-cut. Still, they were consistent and followed suspended matter concentrations, which also decreased offshore (Berryhill 1978).

4
MARINE BENTHIC ENVIRONMENT

with contributions by E. W. BEHRENS, B. B. BERNARD, J. M. BROOKS, P. L. PARKER, R. S. SCALAN, J. K. WINTERS

General Features

The primary topographic features of the STOCS are the deltaic bulge seaward of the Rio Grande, the comparable outline of an ancestral delta of the Colorado-Brazos near the shelf edge seaward of Matagorda Bay, and the broad ramp-like indentation on the outer shelf between the two deltaic bulges. Second order topographic features are the north-to-northeastward trending low ridges, terraces, and low scarps over the ancestral Rio Grande delta, the series of small enclosures associated with a band of irregular topography (e.g., Hospital Rock and Southern Bank) along the ramp between water depths of 64 to 91 m, and the terrace-like area along the outer shelf beginning at the 91-m isobath.

In general, the remainder of the sea floor is characterized by sand-sized sediments on the inner shelf that decrease in abundance seaward. The surficial and near-surface bottom sediments are typically relatively soft and not suitable for bearing heavy structures at shallow depths (Berryhill 1977). Some slumping of the seafloor sediments has been indicated along the periphery of the ancestral Rio Grande delta. Where firm relict sand and soft mud are locally adjacent, sea floor stability is highly variable over short distances. Rapid rates of local sediment deposition or scour have not been observed in this area.

According to results of the 1975 study on the STOCS summarized by Berryhill (1977), sand is transported seaward from the high-energy zone of the innermost shelf. The presence of thin, discrete sand layers in the subsurface sediments to at least 18 km offshore suggests that transport of sand occurs over a relatively short time and is influenced by short-lived events. The encroachment of sand particles onto the Texas shelf from the north suggests a regional southward movement of sediment.

A feature peculiar to the outer continental shelf of the northwestern Gulf of Mexico is the series of pinnacle-like banks or topographic highs rising abruptly from the generally smooth, sediment-covered bottom (Parker and Curray 1956). Two of these, Hospital Rock and Southern Bank, were studied in conjunction with the STOCS survey. In addition, a series of inshore irregularities between 26° north latitude and 27° north latitude (near Transects III and IV, respectively) include scattered rock, shell, and sand banks at 20 to 30 m and 50 to 80 m and a series of inshore elongated troughs and ridges from 10 to 18 m (Mattison 1948). Thayer, La Rocque, and Tunnell (1974) determined some of these features to be of lacustrine origin as well as remains of an earlier, more northerly extension of the Rio Grande delta system. STOCS study transects, however, did not incorporate any of these features.

Sediment Structure

Sediment samples collected at the 25 transect stations 1975–1977 on the south Texas shelf included a wide range of sediment textures from silty clays to muddy sands. Station sites varied enough so that each could be treated as having unique characteristics. More efficient or meaningful comparisons with other data could be made, however, if generalizations were based on groups or gradients of textural data.

The most distinctive sediment group was the outer shelf clays. Textures were graded from finest, best-sorted, and least variable for sediment from the outermost stations (7/IV, 6/III, 3/III, and 3/II) to slightly less well-sorted, siltier, and more variable sediments nearer shore (2/III, 5/III, 5/II, 6/II and 3/I). Collections from the deepest stations generally displayed mean grain sizes of 8.5 to 9.6 ϕ and averaged only 5% sand and 33% silt. Samples from 5/I were very similar to those of this group but more variable. Station 6/I sediment was similar in variability but slightly coarser and less well-sorted with a mean grain size of 7.7 ϕ and 18.5% sand.

Station 4/III sediment was most characteristic of the outer margin (shore face) of the barrier island sand body where variability was low, probably because wave action could constantly maintain a fairly well-sorted texture. Slightly seaward of this zone, sand usually remained predominant but was mixed with considerable amounts (20% to 50%)

of shelf mud (Stations 1/I, 4/I, 1/IV, and 4/IV). The stations on Transect I had fine to very fine sand, while those on Transect IV had much coarser sand and some gravel in the coarse fraction.

The sediment of the rest of the stations on Transect IV (3, 4, 5 and 6) was characteristically very poorly sorted, demonstrating variable mixtures of fine gravel to fine clay. High sample variability suggested that bottom conditions were the least uniform in this environment, with abundant patches of both very clayey and coarse, sandy sediment.

In addition to being related to physical energy intensity and variability, sediment texture variability is also related to variety of source material. Thus, where older sediments are being reworked into more recent material on the Rio Grande delta, there is relatively high variability. The highly variable sediment of Station 5, Transect I, may be similarly related to the ancestral Colorado-Brazos delta sediments at the northern margin of the study area.

Maximum sediment variability was characteristic of a zone just seaward of the boundary of shore face sands (Stations 1/II and 1/III). Fine sand constituted between 10% and 40% of the sediment and was apparently distributed very heterogeneously on scales from centimeters to tens of meters. Adequate statistical sampling in this zone and on the Rio Grande delta would require a larger number of replicates, probably more than have been used in the BLM studies.

The sediment of the last group of stations (2/I, 2/II, and 4/II) represented mid-shelf muds of moderate variability. Although the means for the sediment of these stations were generally in the silt range, silt was almost never predominant, and relative amounts of sand, silt, and clay were extremely variable, with each ranging from 20% to 40%. Consequently, sorting was poor, with a mean of $3.5 \pm 0.3\ \phi$.

The degree of within-station sediment variability did not correlate closely with significant seasonal textural changes. In fact, significant seasonal changes tended to occur in regions that had the most uniform sorting. This suggested that seasonal changes were due to active processes rather than to variability of station sites on successive sampling cruises.

Four stations (1/II, 2/III, 5/IV, and 6/IV) showed significant seasonal changes in sediment texture. The change was an increase in coarseness during the spring accomplished by both an increase in sand and a decrease in clay and little change in silt content. This may have resulted from winnowing of clays and some fine silts during the spring when seasonal winds were at a maximum. The high spatial variability of the Rio Grande delta and the high probability of at least one navigational error having occurred in this region (Station 6/IV), however, made navigational variance a slightly more plausible explanation in this case. Although Station 6/IV sediment was among the group with a high percentage increase in sorting and the additional possibility of seasonal effects, no other indicators suggested a real temporal change at this station, and none was believed to be significant.

The remaining five stations whose sediment showed significant seasonal changes were 1/I, 2/I, 3/I, 3/II, and 4/II. The seasonal changes in sediment at Station 3/II followed the most widespread, significant seasonal changes observed in 1976. Those were spring coarsenings at the outer shelf, clayey sediments accomplished by reduction in the quantity of finest clays ($> 10 \phi$). Sediment of Stations 6/I, 4/II, 5/II, and 6/II also followed this trend, but most of these stations were located without the precision LORAC navigation on the spring cruise when coarsening was observed. Furthermore, the spring coarsening was caused by complex variations in sand, silt, and clay contents rather than just loss of fine clays. Many stations (3/I, 3/III, 5/III, 6/III, and 7/IV) in the outer-shelf group showed no pattern or opposite seasonal trends. Consequently, the trend apparent in the 1976 data of spring coarsening on the outer shelf by winnowing of the finest clays had little support from the 1977 data.

In contrast to sediment of the outer-shelf stations, that of the inner-shelf stations with high sand contents (30% to 80%) showed similar coarsening trends throughout 1977. Although changes at Station 4/III were relatively small, the small intrastation variance made them significant. Coarsening occurred throughout the year, whereas in 1976, spring coarsening was followed by fall fining at this station. Significant spring coarsening also occurred at Station 1/I. All coarsening at inner-shelf stations accompanied increases in sand content and decreases in mud content, resulting possibly from sand deposition, mud erosion, or both. If sand deposition occurred, it would imply a general offshore movement of sand from the barrier shore face. This offshore movement and mud erosion may have resulted from an increase in wave climate. The coarsening effects apparent in the fall seasonal samples may have been related to such an increase resulting from the passage of a hurricane just south of the study area in August preceding the fall sampling cruises. The effect of this event was supported by some fall coarsening at all inner-shelf stations, although sediment of Stations 4/I, 1/IV, and 4/IV did not pass tests of significance.

Station 2/I sediment varied similarly to that of the inner-shelf stations during 1977 in that an increase in sandiness caused the spring texture to be significantly coarser than the texture for the winter samples. There was no significant change, however, between spring and fall samples at this station.

On the other hand, Station 3/I sediment showed spring fining and fall coarsening. These changes apparently resulted from clay deposition in the spring and silt deposition in the fall. These events represented the deeper water equivalents to coarser particle deposition at the inner-shelf stations.

The preceding description of the south Texas shelf sediment structure can be summarized by reference to several sediment variables illustrated geographically in Figure 26 and listed in Table 9. These variables are categorized according to station groupings similar to those listed above but also represent major biotic zones of the benthos that are described in detail in the next chapter. The textural gradients from offshore towards land were as follows: There was a silty (30%) clay of very uniform texture from sample to sample as indicated by mean grain size and its standard deviation (Table 9), a clay that sometimes showed a seasonal tendency to coarsen by the winnowing of finest clays during the early spring at Stations 3/I, 3/II, 3/III, 6/III, and 7/IV. A slightly coarser, more variable silty clay was found at Stations 5/I, 6/I, 5/II, 6/II, and 2/III. These were transition stations between deeper clayey sediments and the silty sediments of the mid shelf. There was quite a variable sand-silt-clay mid-shelf mixture at Stations 2/I, 2/II, and 4/II in the northern part of the study area, and farther landward the most variable inner-shelf, sandy muds occurred at Stations 1/I, 1/II, 1/III, 5/III, and 5/IV. These composed Group 3 of Table 9. A similar group of stations with somewhat more sediment variability, at least partly because of a much coarser sand mode with some gravel, included Stations 2/IV, 3/IV, and 6/IV on the Rio Grande delta (Group 4). Stations 4/I and 1/IV (Group 1) had moderately variable muddy sands near the barrier shore face sand–offshore mud boundary, while Stations 4/III and 4/IV (Group 2) were within the shore face sands where variability became as low as at the outermost stations due to the efficiency of wave action constantly sorting the bottom sediments in shallow water (Table 9). At the inner-shelf stations there was also a suggestion of seasonal coarsening in early spring and a year-long coarsening in 1977 perhaps related to hurricane-generated waves between spring and fall sampling.

Sediment Chemistry

The results of the Delta ^{13}C and total organic carbon analyses for the shelf sediments are summarized in Table 10. A trend of increasing

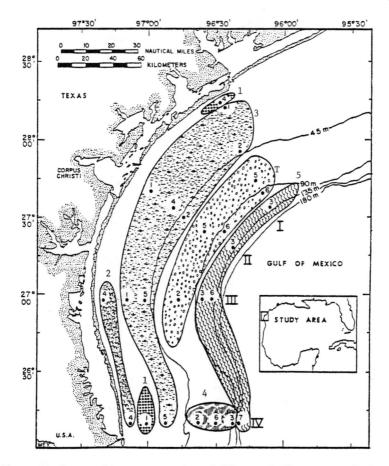

Figure 26. Geographic representation of different shelf zones in which sediment characteristics are relatively similar on the south Texas shelf. Station groups corresponding to Table 9 are shown with arabic numbers. "T" represents the transition station group.

total organic carbon with distance from shore ($P = 0.001$) was very clear. This trend correlated to the percentage of clay in the samples (correlation coefficient = 0.76). There was also a significant change ($P = 0.001$) in Delta ^{13}C with more positive (^{13}C-enriched) values nearer shore. Delta ^{13}C is a measure of carbon-13 enrichment or depletion. A negative value under these circumstances represents an enrichment of ^{13}C and depletion of ^{12}C in relation to a standard with a value of 0. Seagrasses are more ^{13}C enriched than plankton (Calder 1977; Fry

Table 9. Mean and Standard Deviation for Several Sediment Variables for the Station Groupings[a]

Variable	Group 2 (SD)[b]	Group 1 (SD)	Group 3 (SD)	Group 4 (SD)	Transition Stations (SD)	Group 5 (SD)
Depth	15.0 (8.8)	18.5 (0.0)	33.6 (10.2)	67.7 (18.6)	84.6 (13.3)	125.0 (10.3)
Mean grain size	4.10 (0.51)	4.90 (0.42)	7.47 (0.98)	6.68 (1.57)	8.74 (0.70)	9.45 (0.23)
Grain size deviation	2.60 (0.48)	3.57 (0.27)	3.46 (0.27)[c]	3.86 (0.43)	3.15 (0.26)	2.90 (0.15)
Grain size skewness	2.56 (0.84)	1.28 (0.32)	0.37 (0.31)	0.38 (0.46)	−0.03 (0.22)	−0.27 (0.10)
Percentage sand	79.2 (5.1)	65.4 (6.5)	22.1 (14.6)	40.3 (21.5)	8.3 (6.1)	3.2 (2.4)
Percentage silt	9.8 (3.0)	13.3 (3.1)	35.6* (8.2)	19.5 (7.5)	34.8* (5.3)	30.3 (2.1)
Percentage clay	11.0 (4.4)	21.3 (4.5)	42.3 (9.8)	40.2 (14.1)	56.8 (9.2)	66.5 (3.0)
Sand:mud ratio	4.66 (1.2)	2.14 (0.7)	0.36 (0.3)	0.90 (0.6)	0.10 (0.08)	0.04 (0.04)
Silt:clay ratio	1.25 (1.02)	0.65*c (0.19)	0.89 (0.25)	0.49*c (0.04)	0.65* (0.18)	0.46c (0.05)

[a] Station groupings are defined in Figure 26. Analysis of variance indicated significant differences ($P < 0.01$) between all groups.
[b] SD, standard deviation.
[c] Least significant difference results for individual groups not significantly different from each other ($P < 0.05$) are indicated by overlapping horizontal lines or similar superscripts.

1977), and this trend may represent the export of seagrasses from the estuary to the shelf. The bank stations were very uniform.

The rather uniform pattern of Delta ^{13}C and the low values of total organic carbon suggest that petroleum pollution at a fairly gross level could be detected by Delta ^{13}C shifts. If oil of Delta ^{13}C equal to −30 is added to sediment at a level to shift the total organic carbon level from 0.5 to 1.0, then Delta ^{13}C will shift to a value between −20 and −25. Such a total organic carbon shift could go undetected, but such a Delta ^{13}C shift would be easily noted. Even if the oil lost its chemical identity as a hydrocarbon, due to partial oxidation and incorporation into cells, the Delta ^{13}C shift would persist.

Table 10. Summary of Sediment Delta ^{13}C and Percentage Total Organic Carbon

	Near Shore[a]	Mid Shelf[b]	Offshore[c]	Line Average
Transect I				
Winter	−19.92 (0.72)	−20.40 (0.88)	−20.24 (1.02)	−20.18 (0.87)
Spring	−19.58 (0.47)	−20.50 (1.06)	−20.46 (1.04)	−20.18 (0.86)
Fall	−19.24 (0.58)	−19.68 (0.94)	−19.89 (0.56)	−19.60 (0.69)
Yearly	−19.58 (0.58)	−20.20 (0.96)	−20.20 (0.88)	−19.99 (0.81)
Transect II				
Winter	−20.35 (0.70)	−20.35 (0.88)	−20.50 (1.12)	−20.40 (0.90)
Spring	−20.17 (0.93)	−20.38 (0.89)	−20.36 (1.13)	−20.30 (0.98)
Fall	−19.43 (0.82)	−19.65 (1.02)	−20.24 (1.28)	−19.77 (1.04)
Yearly	−19.98 (0.82)	−20.12 (0.93)	−20.36 (1.18)	−20.17 (0.97)
Transect III				
Winter	−19.75 (0.94)	−19.90 (1.02)	−20.10 (0.84)	−19.92 (0.94)
Spring	−19.54 (0.44)	−19.95 (0.97)	−20.32 (1.12)	−19.94 (0.84)
Fall	−18.94 (0.42)	−19.98 (1.01)	−19.88 (1.30)	−19.60 (0.91)
Yearly	−19.40 (0.60)	−19.94 (1.00)	−20.10 (1.08)	−19.82 (0.90)
Transect IV				
Winter	−19.40 (0.73)	−20.10 (0.77)	−20.30 (1.10)	−19.99 (0.90)
Spring	−19.18 (0.28)	−19.75 (0.79)	−19.91 (0.79)	−19.61 (0.62)
Fall	−19.32 (0.21)	−19.99 (1.75)	−20.26 (0.86)	−19.86 (0.94)
Yearly	−19.30 (0.50)	−19.94 (0.82)	−20.16 (0.52)	−19.82 (0.82)
	SB		HR	Bank Average
Bank Stations				
Winter	−20.35 (1.01)		−20.30 (0.70)	−20.32 (0.86)
Spring	−20.26 (1.04)		−20.38 (1.12)	−20.32 (1.08)
Fall	−20.32 (1.03)		−20.19 (1.22)	−20.26 (1.12)
Yearly	−20.31 (1.03)		−20.29 (1.04)	−20.30 (1.04)

[a] Stations 1 and 4 all transects.
[b] Stations 2 and 5 all transects.
[c] Stations 3 and 6 all transects and Station 7, Transect IV.

Statistical analyses provided only weak evidence for temporal and spatial variation of sediment total hydrocarbons. The data suggested sediments contained slightly higher total hydrocarbons in fall, intermediate values in winter, and lower values in spring. Sediment of Transect III stations had the highest concentrations, that of Transect II had middle to high values, and that of Transects I and IV had the lowest values.

We noted statistically significant seasonal effects for the sum of the hydrocarbons between C_{14} and C_{18} (SUM LOW) in 1975 and 1976 data. Seasonal changes in SUM LOW may reflect biological activity and molecular dynamics that take place within the sediment. The high spring and fall values for SUM LOW could result from increased production of these compounds in the water column or at the sediment-water interface. Microorganisms within the sediment consume the added organic matter, including hydrocarbons low in molecular weight, and produce their own characteristic hydrocarbon distribution, which contains a larger percentage of higher carbon number alkanes. Sediment SUM LOW therefore decreases as primary production of hydrocarbons of lower molecular weight decreases.

Long-term temporal changes in sediments were also observed for mid-range hydrocarbons between C_{19} and C_{24} (SUM MID) and the higher ranged hydrocarbons between C_{25} and C_{32} (SUM HI). The data presented in Figure 27 indicate a significant increase in SUM HI ($P = 0.001$) and a concomitant decrease in SUM MID ($P = 0.001$) over the three-year study.

No significant change in OEP HI ($C_{25}-C_{32}$) was observed over the study period despite the tremendous increase in SUM HI ($C_{25}-C_{32}$). OEP MID ($C_{19}-C_{24}$) did, however, show a significant change ($P = 0.001$) as a result of the decrease in SUM MID ($C_{19}-C_{24}$). The lack of change in OEP HI and the changes that did occur in SUM MID and OEP MID suggest that the increase in SUM HI during the study period was due to natural processes rather than the direct addition of petroleum hydrocarbons.

The lack of evidence for the presence of aromatic hydrocarbons in sediments suggested minimal petroleum pollution of STOCS sediments. Petroleum pollution in the form of microtarballs observed in the water column (zooplankton samples) apparently did not contribute a sufficient quantity of petroleum hydrocarbons to sediments to significantly change sediment OEP HI or permit detection of aromatics.

Concentrations of low molecular-weight hydrocarbons in the top few meters of STOCS sediment was generally of microbial origin,

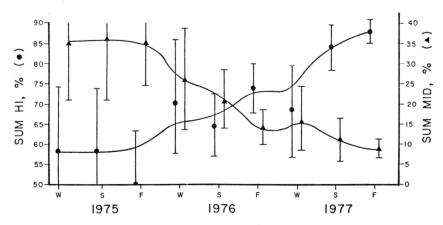

Figure 27. Percentage of sediment n-alkanes in the SUM MID ($C_{19}-C_{24}$) and SUM HI ($C_{25}-C_{32}$) ranges.

as evidenced by the existence of anomalously high methane concentrations in the top sediment layers. Apparently, bacterial production of methane is not restricted to the sulfate-free zone but also occurs within microenvironments in sediments having near-seawater interstitial sulfate concentrations.

Two-meter vertical methane profiles in sediments near shore exhibited maxima ranging from 100 to 500 µl/l (pore water). Figure 28 is a schematic representation of interstitial methane in the upper 4 m of sediment based on samples taken in the STOCS area as compared with slope and abyssal sediments examined independently. The diagram illustrates the disappearance of the maxima as well as the trend of decreasing interstitial methane in an offshore direction. These trends were associated with variations in temperature and microbial activity.

Interstitial concentrations of ethene, ethane, propene, and propane decreased progressively from 160 to 60 nl/l (pore water) in sediments near shore to fairly uniform levels of 80 to 25 nl/l downslope. These trends are illustrated in Figure 29, which shows average concentrations of the four hydrocarbons throughout the cores of Transect I stations, which were sampled independently for comparison of shelf and slope low molecular-weight hydrocarbons.

The trends of the C_2 and C_3 hydrocarbons with distance from shore were similar to the behavior of methane. These patterns suggest that the concentrations of C_2 and C_3 in the top few meters of shelf and

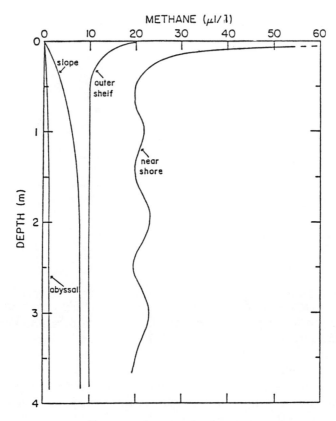

Figure 28. Schematic diagram of methane variations in the upper 4 m of sediment.

slope sediments were microbially supported. Like methane, concentrations of the C_2 and C_3 hydrocarbons are probably controlled by biological oxidation and diffusion into the overlying waters.

The concentrations illustrated in Figure 29 generally represent base line values of the low molecular-weight hydrocarbons on the Texas shelf. One area of anomalously high ethane and propane was found, however, suggesting an input of thermocatalytic gas from the subsurface. Figures 30 and 31 show ethane and propane concentrations with sediment depth at the Transect IV stations compared with those of Transect III stations. Corresponding to the seepage observed in the water column at Transect IV (Chapter 2), sediment low molecular-

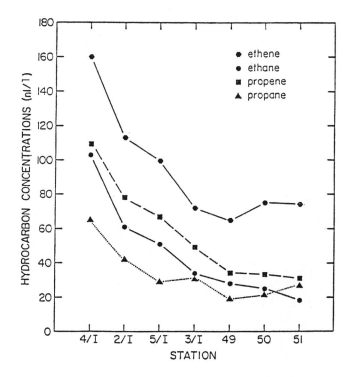

Figure 29. Average concentrations of the interstitial C_2 and C_3 hydrocarbons at Transect I stations. Note that Stations 49–51 were sampled independently, and the data are presented for comparison of shelf (Stations 4/I, 2/I, 5/I, 3/I) and slope (Stations 49, 50, 51) sediments. Stations are arranged by increasing depth.

weight hydrocarbon showed abnormally high concentrations at Stations 3, 4, 6 and 7 along this transect. The stations along Transect III were typical of the normal distribution of low molecular-weight hydrocarbons from biogenic sources in the shelf sediments. Interstitial ethane and propane concentrations generally varied between 20 and 40 nl/l in this area. Ethane and propane concentrations along Transect III (Figures 30 and 31) tended to decrease in an offshore direction with ethane levels generally slightly higher than levels of propane in a manner very similar to Transect I (Figure 29).

The northwestern Gulf of Mexico, including the STOCS area, appears to be free of any significant sediment trace element contamina-

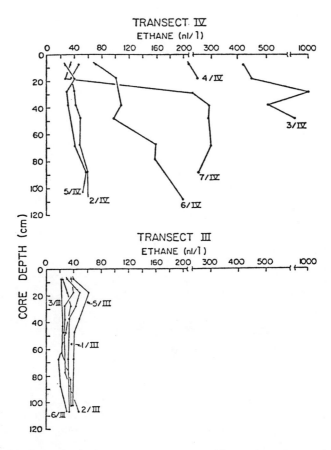

Figure 30. Interstitial ethane concentrations (nl/l pore water) at stations along Transects III and IV.

tion. Metal pollution has been observed in sediments from Corpus Christi Bay (Neff, Foster, and Slowey 1978; Holmes, Slade, and McLerran 1974), the Houston Ship Channel–Galveston Bay area (Hann and Slowey 1972), the Mississippi River delta (Trefry and Presley 1976a), and a few inland waterways (Slowey et al. 1973). There is no evidence of large-scale offshore transport of these contaminants to the outer continental shelf and thus little contamination in shelf sediments (Trefry and Presley 1976b). This situation is not unexpected, especially for the STOCS area, which is not highly industrialized.

Anthropogenic trace elements, along with other materials are

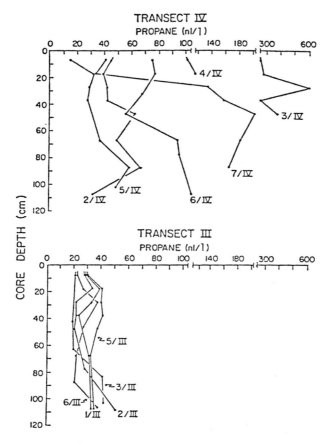

Figure 31. Interstitial propane concentrations (nl/l pore water) at stations along Transects III and IV.

transported to the ocean from continents by freshwater discharges (e.g., sewage outfalls, storm run-off, and river discharge) and atmospheric processes. Direct freshwater discharge into the STOCS region is minimal. The Mississippi and Atchafalaya rivers account for more than 95% of the fresh water entering the northwestern Gulf (Berryhill 1977). The discharge points of these rivers are located approximately 700 km and 500 km, respectively, from the study area. In addition, all rivers on the south Texas coast (except the Rio Grande) discharge into bays and estuaries that are separated from the Gulf by barrier islands.

Without the major industrial areas on the coast and the lack of any direct local riverine impact on the south Texas shelf, high trace metal concentrations in the shelf sediments would not be expected. In general, these trends have been verified by previous work in the northwestern Gulf (Berryhill 1977). Any localized concentration of trace metals in the sediments along the edge of the shelf were attributed to suspected natural gas seepage. Offshore gradients of very low trace metal levels were also shown to be directly related to increases in the clay content of these sediments.

5
BENTHIC BIOTA

with contributions by S. K. ALEXANDER, P. N. BOOTHE, R. W. FLINT,
C. S. GIAM, J. S. HOLLAND, G. NEFF, W. E. PEQUEGNAT,
P. POWELL, N. N. RABALAIS, J. R. SCHWARZ, P. J. SZANISZLO,
C. VENN, D. E. WOHLSCHLAG, R. YOSHIYAMA

One of the major focuses of this multidisciplinary study on the south Texas shelf was characterization of the subtidal benthic habitat from near shore to the shelf slope. As stated by the International Council for the Exploration of the Sea (1978), "A large number of field studies are documented that have as their basis the identification and enumeration of the species occurring in a community, many of which are concerned with the relatively sedentary benthos on the basis that these species will be unable to avoid adverse conditions. Thus, the status of such populations at any point in time is likely to reflect the conditions that prevailed over a relatively long preceding period." The benthos represents an important component of any aquatic ecosystem. Unlike pelagic water masses and associated biota, which are in continual motion, the benthos is relatively stationary and as such serves as a barometer gauging changes that occur in localized areas within the ecosystem. Except for studies of the species with obvious commercial importance, however, the benthos has not received the attention necessary to completely explain the natural variation or to follow the transfer of materials through the communities to which these species belong. It is believed that the benthos serves as one of the essential links in the trophic dynamics of many of the more important fisheries such as the Gulf of Mexico shrimp fishery.

Microbiology

MARINE FUNGI   Fungi are ubiquitous in both terrestrial and aquatic environments. It is somewhat surprising, however, that the predominant genera and species found in sublittoral marine sediments are the same saprobic members of the Fungi Imperfecti that are commonly found in terrestrial habitats (Steele 1967). Despite their documented abundance and the fact that they occur in sediments as viable mycelial filaments (Johnson and Sparrow 1961), the free-living higher fungi have been largely ignored by marine mycologists who have directed their attention to yeasts and less abundant, but uniquely marine, groups of algal parasites and wood-rotting fungi (Jones 1976). The ability of fungi to degrade alkane hydrocarbons (Markovetz, Cazin, and Allen 1968) and aromatic hydrocarbons (Cerniglia et al. 1978) is well documented, but the factors controlling the fungal degradation of crude oil in marine sediments are largely unknown. The study of sediment fungi on the south Texas outer continental shelf is timely because of the rapid increase in petroleum development and production activities in the area.

Marine fungi on the Texas shelf were isolated from sediment samples in 1977. Population densities ranged from a low of five colony-forming units per milliliter sediment (CFU/ml) in winter samples from Station 3/I to a high of 1600 CFU/ml in the fall samples from 3/II (Table 11). The average for the year in the study area was 236 CFU/ml. There was a progression toward larger fungal populations beginning with the late-winter low and ending with a significant ($P < 0.03$) increase in the fall. An exception to this trend was seen at the deep station on Transect I where fungal abundance was much greater in the spring than in the fall. The annual pattern of increasing numbers of fungi through the fall period was paralleled by an increase in generic richness, an index of community diversity. The average number of genera per station increased from 5.8 in the winter and 7.0 in the spring to 10.3 in the fall.

When the abundance of fungi capable of degrading petroleum products (as measured by assaying the growth of pure isolates on crude oil) was compared to that of nonhydrocarbon degraders, 52% of the 83 benthic isolates tested were observed as capable of assimilating crude oil (Table 12). It was clear that a greater proportion of fungi from the shallow stations than those from intermediate depth stations were capable of degrading oil. Oil degradation potential decreased offshore.

Crude oil stimulated the growth of benthic fungi (Figure 32). The addition of south Louisiana crude oil (SLCO) to fall benthic sediment samples resulted, after 45 days, in an average 7-fold increase in fungal abundance at the 0.5% (volume/volume) oil level and a 3.6-fold increase at the 0.1% oil level relative to the control. There was, how-

Table 11. Fungal Abundance in Surficial Sediments by Depth and Season[a]

Depth	Station/Transect	Fungal Abundance (CFU/ml)[b]			Mean (CFU/ml)
		Winter	Spring	Fall	
Shallow	1 / II	110	33	200	98
	1/IV	16	30	200	
Intermediate	2/II	11	20	910	248
	2/III	12	83	450	
Deep	3/I	5	350	160	359
	3/II	21	15	1600	
Mean		29 ± 40	89 ± 130	587 ± 571	

[a] One-way ANOVA with season as independent variable ($F = 4.8331$; $degrees\ of\ freedom = 2,13$; $P = 0.024$).
[b] Colony-forming units/ml wet sediment.

Table 12. Growth of Fungal Isolates in Crude Oil by Station and Depth[a]

Depth	Station/Transect	No. of Isolates Showing Growth	No. of Isolates Showing No Growth
Shallow	1/II	11	6
	1/IV	13	9
Intermediate	2/II	5	9
	2/III	3	7
Deep	3/I	3	3
	3/II	8	6

[a] Fungi isolated from benthic sediments on nonselective medium. By depth, $X^2 = 4.916$; $degrees\ of\ freedom = 2$; $(0.05 < P < 0.1)$.

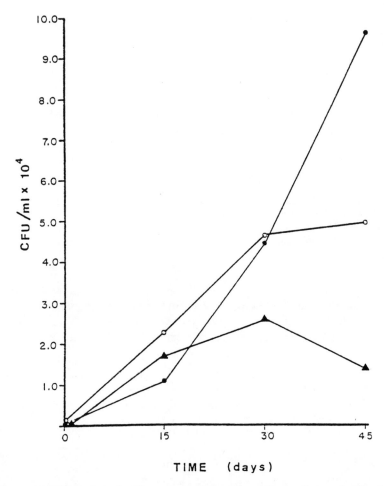

Figure 32. Effect of crude oil concentration on fungal growth in natural mixed cultures of STOCS benthic sediments diluted (1:5) with artificial seawater. The numbers are the mean values for all samples from the fall cruise. Symbols are (●), 0.5% oil (SLCO); (○), 0.1% oil (SLCO); (▲) control to which no oil was added.

ever, an initial inhibition of the natural mixed fungal populations in the 0.5% treatment. This initial toxicity was also seen in experiments with pure cultures of *Candida diddensii* in which prestarved inoculum and low nutrient conditions duplicated as nearly as possible STOCS ecosystem conditions. Severe toxicity was observed in all cases with early survival rates ranging from 0% to 3% of the no-oil control. Maximum toxicity occurred between the third and sixth days with recovery and significant stimulation taking place by the 22nd day.

The abundance of fungi in the STOCS benthic system appears to be controlled by two factors: (1) the replenishment of inoculum from the water column, a seasonal phenomena, and (2) the availability of organic carbon, a site-specific variable. In general, the genera of fungi observed during the study were those whose spores are usually most abundant in the air over the adjacent land masses. The species most frequently encountered in sediment samples during this study were *Cladosporium cladosporioides* (Freson.) deVries, *Penicillium citrinum* Thom, *Aspergillus flavus* var. *columnaris* Raper and Fennel, *Aspergillus sydowi* (Bain and Sart.) Thom and Church, *Fusarium ventricosum* Appel and Wollenweber, and *F. moniliforme* var. *subglutanans* Wr. and Reink. The terrestrial origin of these genera was also suggested by higher abundances observed at the stations near shore (Table 11).

The large increase in benthic fungi isolated from the sediments in the fall can be explained by the early fall arrival of spores suspended throughout the summer at the thermocline-pycnocline zone following their deposition in the water column during late winter and spring. As the atmospheric spore load in Texas is reaching its annual maximum (Chapman 1979), the last continental air masses of spring are moving out over the Gulf of Mexico off Corpus Christi (late April or early May) (Orton 1964). Until fall the area is covered by maritime air masses. These conditions are reflected in the abundance of fungi in STOCS near-surface waters (Szaniszlo 1979). During March and April of 1977 fungi were uniformly very abundant with monthly averages of 40,000 and 16,000 CFU/l compared with only 13 CFU/l in July, 4 CFU/l in August, 21 CFU/l in November, and 4 CFU/l in December.

The number of colony-forming units of benthic marine fungi observed in the fall samples was directly correlated to the total organic carbon concentrations of these sediments ($r = 0.843$). Indirect evidence also existed from the observations of this study suggesting that fungal mineralization of the organic material during winter, spring, and summer controlled the peak abundance of fungi observed during the fall. Fungi appeared to be short-lived in the STOCS sediments in which available carbon was the limiting factor. Over half of the benthic fungi tested were able to assimilate south Louisiana crude oil to overcome carbon limitation.

Since organic carbon, and not nitrogen or phosphorus, limited fungal abundance in the STOCS ecosystem, it is reasonable to presume that at least some fungal oxidation of intrusive petroleum would occur anywhere in the area. Greater activity, however, would be expected inshore in coarse sediments subject to freshwater outwash high in nutrients.

MARINE BACTERIA   Aerobic heterotrophic bacteria ranged from 4.6×10^4/ml wet sediment to 1.3×10^6/ml wet sediment. Analysis of variance indicated a significant ($P < 0.01$) seasonal difference in benthic bacterial populations, with highest numbers during spring and lowest during winter (Table 13). There was no significant difference between transects. There was, however, on each transect a significant difference between stations, with highest populations at Station 1, decreasing with increasing depth (Figure 33). Mean populations of benthic bacteria at Stations 1, 2, and 3 (all transects and seasons) were 7.9, 4.3, and 2.2×10^5/ml wet sediment, respectively. The variation by station accounted for 47% of the total variance in benthic bacteria. The only deviation from this distribution was on Transect IV, where Station 3 contained an unusually high number of bacteria during the spring and fall.

Rates of organic carbon input to the sediments during the spring are expected to be greater than at any other time during the year because of peak productivity measures in the overlying water column.

Table 13. Summary of Benthic Bacterial Populations During 1977

Type	Season	No. of Samples	Mean ± SD
Aerobic heterotrophic bacteria	Winter	24	$40.2 \pm 28.0 \times 10^4$
(number/ml wet sediment)	Spring	24	$55.4 \pm 41.6 \times 10^4$
	Fall	24	$47.8 \pm 31.2 \times 10^4$
Hydrocarbon degrading	Winter	24	$2.7 \pm 3.8 \times 10^3$
bacteria (number/ml wet	Spring	24	$3.1 \pm 5.1 \times 10^3$
sediment)	Fall	24	$23.5 \pm 35.8 \times 10^3$
Percentage hydrocarbon	Winter	24	0.6 ± 0.5
degrading bacteria	Spring	24	0.5 ± 0.4
	Fall	24	4.8 ± 6.1

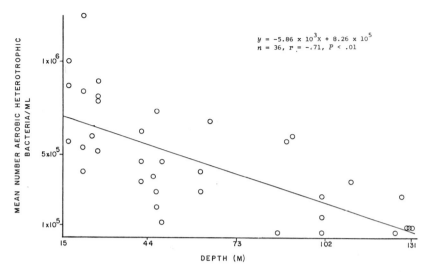

Figure 33. Relation between aerobic heterotrophic bacteria of sediment and bottom depth.

Benthic bacteria appear to have responded to this high input during the spring, since this is the period of maximal populations. Temperature may also affect the seasonal distribution of benthic bacteria. Benthic bacterial populations were lowest during the winter, corresponding to seasonal low sediment temperatures.

Hydrocarbon-degrading bacteria were isolated from all 72 samples collected during the study. Populations ranged from 8.0×10/ml sediment to 1.1×10^5/ml sediment and were significantly correlated with total alkanes of the sediment (Figure 34). Analysis of variance demonstrated a significant ($P < 0.01$) seasonal variation in the number of hydrocarbon-degrading bacteria, with highest populations during fall and lowest during winter (Table 13). There was a significant ($P < 0.01$) difference between transects, with greatest concentrations on Transect I during winter and spring and on Transect IV during fall. Hydrocarbon-degrading bacteria were also significantly ($P < 0.01$) greater on all transects at Station 1, decreasing with increasing depth. The mean number of hydrocarbon-degrading bacteria at Stations 1, 2, and 3 (all transects and seasons) was 17.3, 9.3, and 2.6×10^3/ml wet sediment, respectively.

Benthic bacteria are capable of degrading all n-alkanes from C_{14} to C_{32} but exhibit a preference for the lower ranged high molecular-

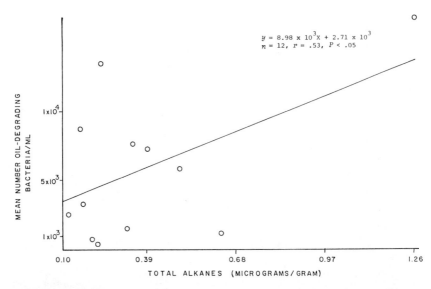

Figure 34. Relation between number of oil-degrading bacteria of sediment and total alkanes.

weight hydrocarbons (C_{14} to C_{20}). Spatial variations in biodegradation of oil were examined for each season. No significant spatial variations occurred in the winter, probably due to consistently low rates of biodegradation (from 0% to 16.78%). During the spring, there were significantly ($P < 0.01$) higher biodegradation potentials at Station 1, decreasing with increasing depth. There was also a significant ($P < 0.05$) difference between transects, with lowest potentials on Transect III. During the fall, there were no significant spatial variations in biodegradation potential. The mean percentage of biodegradation of oil during the spring and fall was significantly correlated to the mean number of hydrocarbon-degrading bacteria ($r = 0.66$ and $r = 0.73$, respectively).

Oil significantly ($P < 0.01$) stimulated the growth of total aerobic heterotrophic bacteria at the majority of stations during the three seasons. Growth stimulation by SLCO occurred after one week and continued through eight weeks. Significant growth inhibition by oil was not observed. The number of hydrocarbon-degrading bacteria of sediment was also significantly ($P < 0.05$) increased by the addition of oil. Stimulation of hydrocarbon-degrading bacteria by oil was recorded after two days and continued through eight weeks.

In conclusion, two study findings suggest that growth of hydrocarbon-degrading bacteria may be a useful indicator of sediment hydrocarbons in the STOCS area: (1) the number and percentage of hydrocarbon-degrading bacteria were significantly correlated with total alkanes of the sediment, and (2) the addition of oil to the sediment increased the number and percentage of hydrocarbon-degrading bacteria after two days.

Meiofauna

The meiofauna has been largely ignored until the last three decades. The importance of the economic aspects of macrobenthos to fisheries and the function of microorganisms in converting organic material into usable energy in a food chain have received the majority of research attention, the former to a considerable extent. Meiobenthic work has, until recently, been confined to study of species composition, diversity, and density or to detailed examination of a particular group. In the last 10 years increasing attention has been focused on the ecology of the marine meiobenthos and its trophic interactions.

"Meiobenthos" was first used by Mare in 1942 to characterize benthic fauna of intermediate size, such as small crustacea, small polychaetes and lamellibranchs, nematodes, and foraminifera. The distinction was to separate the intermediate-sized metazoans from larger macrofauna of the bottom and the microbenthos—protozoa (excluding foraminifera), diatoms, and bacteria. This arbitrary size definition, usually accepted as animals that pass through a 0.5-mm sieve but are retained on a sieve with mesh smaller than 0.1 mm (Coull 1973), may include representatives of the young of the macrofauna (temporary meiobenthos) but are more commonly accepted only in terms of species that even at the adult stage fit into the stated size category and fit certain taxonomic categories, the permanent meiobenthos (McIntyre 1969).

The meiobenthos designation is considered purely statistical (McIntyre 1969), with no clear-cut distinction between the macrofaunal and meiofaunal components. Further delineation as "permanent," however, provides a more operational definition in terms of sampling methods and a natural grouping with certain biological characteristics (McIntyre 1969). McIntyre (1969) further defined meiofauna as an "assemblage of small metazoans which differ from their larger counterparts (macrofauna) in their reproductive capacity and general metabolism, as well as in the ecological niche they fill." The two components may (1) compete with each other for resources (McIntyre 1964 and 1969), (2) lack significant interaction between each other, or (3) operate independently of each other while being controlled by different environmental factors (McIntyre 1974). Study into the trophic relations and microecology of meiofauna indicates that it is as intricately entrenched in the integrated marine food web as the macrofauna but differs in activities and requirements.

During both 1976 and 1977 meiofaunal populations diminished with increasing depth on the Texas shelf (Figure 35). Consistently, Transect IV supported the highest populations inshore and Transect II the lowest. Populations of the deepest station of Transect II were

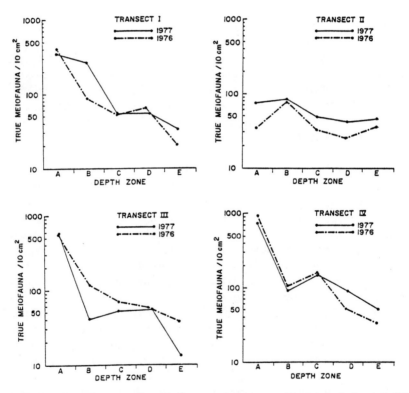

Figure 35. Distribution of permanent meiofauna on Transects I through IV during the winter, spring, and fall sampling periods by depth zone for 1976 and 1977. Points are means of populations for three seasonal sampling periods plotted logarithmically. Depth zones are A (0–30 m), B (30–60 m), C (60–90 m), D (90–120 m), and E (> 120 m).

almost as great as those of the shallowest station. In contrast, for the other three transects, populations of the deepest stations were only a small percentage of those of the shallowest stations.

Pequegnat and Sikora (1977) reported that monthly sampling was necessary to define temporal variability of meiofaunal populations. This was best shown by nematodes on Transect II, which was the only transect sampled more than three times during the year (Figure 36). There were population peaks in March, July–August, and November. Population peaks were much greater inshore than offshore. Figure 36 also shows that the March 1976 inshore population was

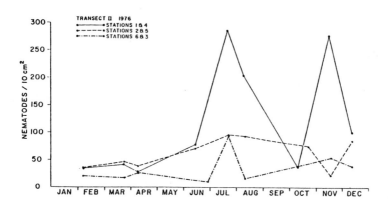

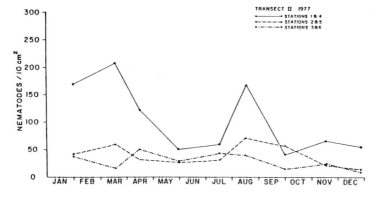

Figure 36. Monthly distribution of nematodes at inshore stations (1 and 4), mid-depth stations (2 and 5), and offshore stations (6 and 3) of Transect II during 1976 and 1977.

very small and November's was large, followed by a very large March 1977 population and a reduced November population.

Nematodes were the most abundant meiofaunal taxa observed, averaging 92.6% of the total abundance of the permanent meiofauna (Table 14). Transect II for the year 1976 averaged 86.9% nematodes and was the only case of a transect averaging less than 90% in the two years. The numerically dominant nematodes are listed in Table 14. *Sabatieria* occurred very commonly in sandy silts and muds, regardless of water depth. *Laimella* also occurred primarily in sandier sediments. There was a marked increase in nematodes when the sand content of the sediment was 60% or more by weight. The high percentage of nematodes in the samples was comparable to that found in muddy continental shelf areas in the Kerguelen Islands (Bovée and Soyer 1977), off Massachusetts (Wigley and McIntyre 1964) and also at an 18-m mud station in Buzzards Bay, Massachusetts described by Wieser (1960).

The second most abundant taxon in the STOCS study was Harpacticoida. Harpacticoid populations were proportionately much smaller than those of the nematodes (Table 14). Inverse of that of the nematodes, the proportion of harpacticoids was somewhat higher at Hospital Rock and Southern Bank, averaging 8.5% of the permanent meiofauna for both banks together over 1976 and 1977. The high percentage of harpacticoids may have been more a result of the very reduced total meiofauna populations at those stations, thereby increasing the proportional effect of an occasional occurrence, rather than a true indication of increased harpacticoid abundance. Numbers of harpacticoids taken at the transect stations ranged from 0 to 97.4 individuals/10 cm^2 in 1976 and from 0 to 59.3 individuals/10 cm^2 in 1977. The mean for all stations was 662.0 individuals/10 cm^2 in 1976 and 458.7 individuals/10 cm^2 in 1977.

Kinorhyncha were not very abundant, averaging about 0.5% of the taxa over all transect stations for both years (Table 14). They were even less abundant at the two bank stations, totaling 27 kinorhynchs from all stations and sampling periods in 1976 and 24 in 1977.

Polychaeta was the second most abundant taxon of the total meiofauna (excluding the Foraminiferida and Protozoa) on Transects I through IV (Table 14), totaling 3593 individuals collected over the two years of the study. As with the nematodes and the mean permanent meiofauna, the Polychaeta highest abundances were in the shallow zone (0 to 30 meters), with numbers decreasing at the offshore stations. Abundances ranged from 0 to 115.9 individuals/10 cm^2 with a mean of 7.0 individuals/10 cm^2 in 1976 and from 0 to 47.6 individuals

Table 14. Major Taxa of the Meiobenthos

Location/classification	Nematoda (% true meiofauna)	Harpacticoid (% true meiofauna)	Kinorhyncha (% true meiofauna)	Polychaeta (% total meiofauna)[a]
All Stations				
1976—Mean	94.5	3.1	0.4	4.6
Range	100.0–61.6	25.7–0.0	3.1–0.0	25.0–0.0
1977—Mean	90.7	4.1	0.6	4.1
Range	100.0–55.9	15.4–0.0	7.3–0.0	16.3–0.0
Both Years				
Mean	92.6	3.6	0.5	4.3
Banks				
Both Years				
Mean (%)	84.1	8.5	0.6	5.4
Range (%)	100.0–40.9	36.0–0.0	3.3–0.0	16.7–0.0
Genera or Species[b]	Sabatieria	Haloschizopera	Echinoderes	Paraonis gracilis
	Theristus	Enhydrosoma	Pycnophyes	Tharyx setigera
	Halalaimus	Pseudobradya	Semnoderes	Mediomastus californiensis
	Dorylaimopsis	Ameira	Trachydemus	Aedicira belgiacae
	Neotonchus	Ectinosoma	Centroderes	Protodorvillea sp. A
	Terschellingia	Typhlamphiascus		Cossura delta
	Synonchiella	Robertgurneya		Aricidea cerruti
	Viscosia	Halectinosoma		Sigambra tentaculata
	Lainella	Thompsonula		Prionospio cristata
	Ptycholaimellus	Apodopsyllus		
		Leptopsyllus		
		Stenhelia		

[a] Percentage of total meiofauna, excluding Foraminiferida and Protozoa.
[b] Genera are listed in order of numerical abundance.

/10 cm² with a mean of 5.8/10 cm² in 1977. Numbers of polychaetes averaged less for the bank stations than for the transects.

Meiofauna in general are similar to macrofauna in that they are not a homogeneous group. They employ many of the same varied feeding mechanisms as the heterotrophic macrofauna and are subsurface specialized deposit feeders, microbial consumers, nonselective subsurface deposit feeders, and predators (Gerlach 1978; McIntyre 1964). Still others are highly specialized and have adapted physiologically, such as those in a sulfide community in which the dominant forms are ciliates and a few metazoans capable of existing in a reducing environment by employing surface existence or a very narrow vertical range (Coull 1973).

Meiofauna share similar habitats with macrofauna, both being found in all marine ecosystems, estuaries, sandy beaches, subtidal muds, and the deep sea. Macrofauna show a pattern of distribution similar to that of the meiofauna influenced primarily by sediment parameters (Wieser 1960) with preference for sandy or silty sediment. In contrast, life histories of meiobenthos are very diversified, probably no less so than those of different macrofauna (Gerlach 1971). With constant numbers spawning all year in some habitats and not restricted by season, the resultant productivity may equal or excel that of macrofauna (McIntyre 1964).

Macrofauna

INFAUNA   Since Petersen published his work (1913 and 1918), investigators have delineated benthic communities in relation to environmental parameters such as hydrologic variables (Molander 1928), physical properties of the bottom sediments (Jones 1950), and biological adaptations derived from species interactions in relatively stable environments (Sanders 1968). Community distributions have been examined in a number of different aquatic environments in recent years. These studies have found that the benthos varies considerably in space due to the general heterogeneity of aquatic systems and the tendency towards patchiness in the benthic fauna.

The development of a large multidisciplinary research program in a little-studied subtropical area of the Gulf of Mexico off the south Texas coast provided the opportunity to contrast benthic community structure and factors influencing this structure with other continental shelf ecosystems. The south Texas shelf is comprised of much siltier, less stable sediments than other shelves such as the middle Atlantic region, which is characterized by sandier sediments out to greater depths on the shelf (Boesch 1979). The outer Texas shelf can also be considered a true soft-bottom environment because unlike other shelves of the eastern and southern Gulf or south Atlantic or Pacific, there are very few reef areas or extensive banks with their influential biogeographic effects, that is, "islands in a sea of mud." Additionally, with the pressure of extensive energy exploitation slated for the near future on the south Texas shelf, it is imperative to document the species assemblages of the benthos in a relatively pristine habitat. Although pristine, this habitat is one which would probably be most directly affected should a major environmental disturbance occur (e.g., oil well blowout) and one which is a direct supportive element to many of the regional fisheries such as shrimp.

Ordination analysis of the infaunal species composition for each of the 25 collection sites indicated that 73% of the total variation between sites was accounted for by the first and second coordinates. The third coordinate only accounted for an additional 4% variation and showed no meaningful trends. Therefore, all emphasis was placed on the first two coordinates (x and y axes).

In order to objectively define community differences within each collection site and station, scores from community ordination were evaluated by the least significant difference (LSD) multiple range test. Both coordinate mean scores of the six collection periods were compared for each station. The results showed that the first ordination coordinate was able to significantly delineate ($P < 0.05$) four station groupings, Groups I, II, III, and V (Figure 37). Station Group I consisted of Stations 4/I and 1/IV, and Group II was composed of collections from Stations 4/III and 4/IV.

Station Group III was defined by the largest number of collection sites and included mid-depth stations. According to the LSD results for the second ordination coordinate, three collection sites on Transect IV significantly ($P < 0.05$) differed from the other sites in Station Group III and were thus considered a group within themselves (Station Group IV). Station Group V was composed of the five deepest stations that showed consistently low scores for both the first and second ordination coordinates. A group of five stations, 5/I, 6/I, 5/II, 6/II, and 2/III, did not show a significant difference from most sites in Station Group III or V according to first coordinate mean and were not further differentiated by the second coordinate; therefore, these stations were defined as being in a transition zone between the mid-shelf communities (Station Group III) and the deep water communities (Station Group V). Cluster analyses on the same data showed similar results. These station groupings are also illustrated in Figure

Benthic Biota

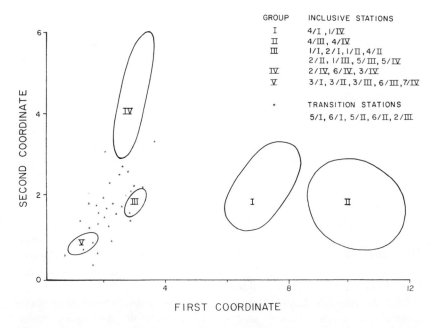

Figure 37. Results of benthic infauna community ordination. The group 95% confidence ellipses are shown for the first and second coordinate from ordination. Asterisks represent transition stations.

26, which suggests that the infaunal station groupings and benthic habitat environmental variables showed similar trends.

The community variables of species numbers, infauna density, species diversity, and equitability exhibited trends that were consistent with the community ordination results for the station groups (Figure 38). The number of species was consistently the highest at the shallow stations (Groups I and II), with a sufficient drop for Group III sites. Organism density was also greatest for the shallowest sites with decreases in deeper waters on the shelf. High species numbers and infaunal densities resulted in high species diversity for the shallow stations (Groups I and II). The highest diversities, however, were measured for Station Group IV. Equitability showed an increase in the offshore direction, indicating that although the shallow collection sites were high in species numbers and densities, these sites were characterized by a few dominant fauna in contrast to more evenly distributed population densities for the offshore species assemblages (Groups IV and V). Results of one-way ANOVA indicated there was

Benthic Biota [99]

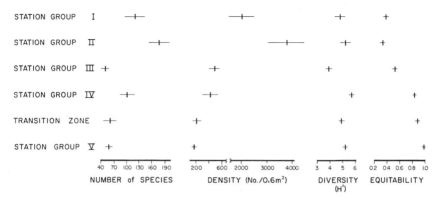

Figure 38. Plots of station group means for infaunal community characteristics. Station groups defined from Figure 37. Bars represent 95% confidence intervals.

a general significant difference ($P < 0.01$) between station groups for all four community structure measures. LSD range tests, however, showed no difference between the transition zone and at least one of the station groups bordering this zone (III or V) for each of the four variables presented in Figure 38, emphasizing the transitional nature of these fauna.

Inverse community ordination (R mode), identifying the species characteristic of collection sites, aided in the interpretation of specific faunal assemblages that described the shelf station groups highlighted above. Fifty-eight infaunal species that showed a minimum of at least 1% abundance at a station during the study period were identified by this ordination analysis (Figure 39). These 58 species made up six groups that were coincident with one or more of the station groups. Species Group I represented shallow water fauna, while Species Groups II and III consisted of infauna showing shallow to mid-shelf distributions. Species Group IV was comprised almost exclusively of deep water infauna. Groups V and VI were comprised of infaunal species relatively ubiquitous over the south Texas shelf. Most major taxonomic classes (i.e., polychaetes, molluscs, crustaceans) were represented by these faunal groupings, but polychaetes were by far the dominant taxa of the shelf benthos. As illustrated by monthly sampling on Transect II (Table 15), there was a decrease in polychaete domination in the offshore direction, but polychaetes still made up the majority of the community at all stations.

The transition stations again illustrated why they were called such,

[100] Benthic Biota

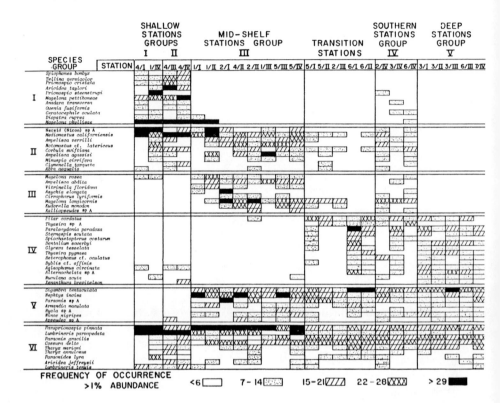

Figure 39. Infaunal species frequency of occurrence for the STOCS sampling stations presented according to station group defined from Figure 37. Species groups are also indicated by roman numerals.

showing the abundance of species characteristic of both mid-shelf stations and deep stations (Figure 39). The frequency of species occurrences within Station Group IV suggested that this area may be unique on the Texas shelf. Although in deeper water, these sites supported fauna characteristic of shallower stations (e.g., *Prionospio steenstrupi* and *Ceratocephale oculata*) as well as fauna representative of collection sites within their depth range. Figure 39 further illustrates two points stressed in Figure 38. First, the occurrence of more species in Station Group IV contributed to the high species diversity of this group on the shelf. Secondly, the proportionately greater species diversity with higher frequencies of occurrence at the shallower stations hinted the dominance of some of these organisms of the shal-

Table 15. Mean Percentage Composition of Major Faunal Groups at Six Stations on Transect II Over Study Duration

Faunal Group	Station					
	1	2	3	4	5	6
Polychaeta	81.6	67.7	67.4	49.6	53.2	52.6
Mollusca	2.3	6.0	10.2	34.2	20.0	14.4
Gastropoda	1.9	2.9	2.3	3.5	2.9	1.9
Pelecypoda	0.3	3.1	7.7	29.1	15.1	9.0
Crustacea	10.6	19.7	14.3	8.4	16.0	21.5
Ostracoda	—	—	0.1	0.4	1.6	8.0
Isopoda	—	—	—	0.3	5.2	3.0
Amphipoda	9.1	10.7	6.2	3.7	6.7	5.6

lower shelf waters (i.e., *Magelona phyllisae*, *Paraprionospio pinnata*, and *Lumbrineris verrilli*).

Assemblages of organisms are rarely present as discrete groups with clear-cut boundaries as evidenced by the need for a transition community in the observations presented above. Groupings of organisms into communities must therefore be inferred from consideration of the interactions of the fauna with various environmental factors. This grouping is also evident in the similar patterns observed between the infaunal station groupings (Figure 37) and the station groupings for the sediment characteristics (Chapter 4, Figure 26 and Table 9). Multivariate discriminant analysis was used to aid in identifying discriminating environmental variables for the infaunal station groupings and also to test the null hypothesis that there was no environmental difference between these groups.

Figure 40 illustrates the position of station group mean discriminant function scores with their 95% confidence ellipses in a two-dimensional space defined by the first and second functions (Figure 40A) and the first and third functions (Figure 40B). Also indicated on each plot are the individual transition station scores for each collection period. From a suite of 13 original environmental variables, four variables proved to be good discriminators of the infaunal station groupings (Figure 41).

The first and second discriminant functions accounted for 94.7% of the variation between station groups and were both significant ($P < 0.01$) in discriminating between groups as indicated by their chi-square values (Figure 40A). Approximately 80% of the variation in the first function was accounted for by the environmental variable of

[102] Benthic Biota

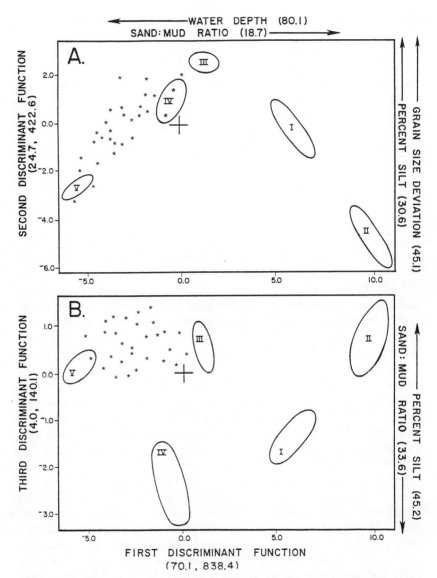

Figure 40. Results from discriminant analysis of environmental variables according to infaunal station groupings (Figure 37). The 95% confidence ellipses for each group plus the individual points for the transition stations (asterisks) are plotted for the first and second (*A*) and first and third (*B*) discriminant scores. Environmental variables responsible for group separation are shown with their explained percentage variation in parentheses.

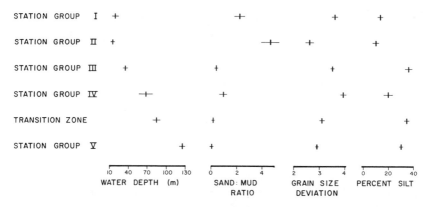

Figure 41. Station group means and 95% confidence limits for environmental variables responsible for station group separation in Figure 40. Bars represent 95% confidence intervals.

water depth. The sediment variable, sand : mud ratio, accounted for an additional 18.7% of the first function's variation. Figure 41 illustrates that water depth was able to significantly differentiate the shallow stations (Groups I and II) from the mid-shelf stations (Groups III and IV) and the mid-shelf stations from the deep stations (Group V). The sand : mud ratio was further able to differentiate Station Groups I and II from each other.

On the second discriminant function (Figure 40A), sediment grain size deviation further distinguished differences between Groups I and II (Figure 41) as well as showing a more subtle but significant split between Groups III and IV. The sediment variable, percent silt, however, showed the strongest differentiation between the latter groups.

The third discriminant function (Figure 40B) was also significant ($P < 0.05$) and accounted for an additional 4% in variation between station groups. This function further illustrated the discriminating power of percent silt in not only differentiating Groups III and IV but also Group IV from the transition stations, which were in the same general depth range (Figure 41).

The overall chi-square derived from the general Mahalanobis distance squared was 640.9 for the discriminant analysis of station groups according to environmental variables. This chi-square was highly significant ($P < 0.001$) and suggested that the null hypothesis of no environmental difference between groups be rejected. It was assumed, therefore, that there was very little probability the station groups could have been formed by chance but that the separation between them was real. These results confirmed the biological model and suggested some of the variables potentially influential in structuring the infaunal communities along the shelf depth gradient.

Sediment properties appeared to be relatively important in terms of community structure patterns, according to Figures 40 and 41. Although these properties are mildly correlated with water depth—the most powerful discriminating variable observed—there are other factors also related to water depth that must be considered in the interpretation of results from the study. These factors include the amount of food available to the benthos and bottom water environmental variability along the depth gradient as characterized by surface chlorophyll *a* concentrations and the standard deviation measure of temperature and salinity. Data presented in previous chapters indicate that these variables decrease with increasing water depth on the shelf. Chlorophyll *a* concentrations were highest and also most variable in shallow waters (Chapter 3) in which the highest densities of infauna were observed. Lower concentrations of primary producers, whose abundances were less variable throughout the study interval at deeper stations, were associated with lower densities of infauna and more evenly distributed population numbers (equitability) within these assemblages (Figure 38).

Temperature and salinity were both most variable at the shallower collection sites, with variability decreasing as water depth increased (Chapter 2). This pattern implied that the shallower benthic habitat was much more variable and less predictable in terms of environmental changes and, therefore, conducive to dominance by a few fauna. This variability of the shallow shelf was further verified by the fluctuations of chlorophyll *a* representing a food source to the benthos through the detrital pool. In addition to the influential effects of certain sediment characteristics on benthic community structure, gradational features of a food source to the benthos and variability in the bottom water environment were also suspect as potential causes of the different faunal patterns observed.

Other benthic marine systems investigated have been shown to be typically gradational in space with respect to sediment and other environmental variables (e.g., Day, Field, and Montgomery 1971; Field 1971; Boesch 1973; Glemarec 1973). Closely correlated to these environmental changes are changes in macroinfaunal communities. According to the observations presented above, sediment composition plays an important role in structuring the benthos. Superimposed on the mechanics that the substrate pose on the benthic infauna, however, are factors involved in producing variability both to a food source of the benthos and the overlying hydrologic environment. These environmental aspects come together to produce a very complex association between the Gulf of Mexico benthos and the habitat in which the organisms live.

According to Glemarec (1973), the nature of the sediments is of prime importance to the settlement of most invertebrate larvae and the resultant composition of communities. He extends his definition of spatial stages of the benthos, however, to include the effects of variations in bottom water temperature and cites examples from Jones (1950) and Lie (1967). Glemarec concludes that the environmental properties that permit a distinction between faunal assemblages are different depending upon whether the assemblages are in shallow or deep water.

Significant variability in shallow waters combines with coarse ill-sorted sediments to provide an unstable habitat. This habitat is characterized by many different fauna with few exhibiting dominant abundance (low evenness). In contrast, another habitat with coarse sediments (Station Group IV) exhibits the most diverse fauna observed during the study period. These sites, in addition to having a very heterogeneous sediment structure, are characterized by very stable hydrologic variables as well as a more predictable food source.

As Sanders (1968) and McCall (1977) illustrated, in a marine habitat subjected to continual local disturbances and harsh environmental variables, as Texas inner-shelf waters are, a few highly specialized species—opportunists according to Grassle and Grassle (1974)—are present in large numbers. These species are able to invade new areas voided of fauna by a local disturbance (i.e., currents) and to maintain their large population sizes because of the abundant food sources and unpredictability of the bottom water environment, which may be occasionally disturbed by storm currents. In contrast, deeper shelf habitats exhibit less bottom water variability, and sediment characteristics become the key to faunal distribution. This is evident in the faunal changes between the mid-depth stations (Group III) where the silt content is high and the deep water stations where clay is the more dominant sediment component.

There was a variable sand-silt-clay mid-shelf mixture observed at most stations between water depths of 20 and 50 m (Station Group III) with silt the dominant component. These stations generally showed a sand : mud ratio of 0.3 to 0.5, much lower than the shallow study sites. The percentage of silt was also a major discriminating variable separating Station Groups III and IV (Figure 41). Group III exhibited the lowest number of infaunal species on the shelf while supporting population densities second only to the shallow sites. Associated with these community variables were low measures for both species diversity and equitability, suggesting that these species assemblages were dominated by a few fauna with high densities.

Rhoads (1974) stated that silty sediments present a difficult environment to which few species can adapt. Not only are the ecological niches decreased by a more homogeneous substrate (Ward 1975), but the vulnerability of particle sizes to bottom water currents is greater. This can produce a relatively unstable substrate for infauna inhabitants. A good example of the instability of this particular area is the sediment resuspension associated with the nepheloid layer that occurs frequently during the year (Kamykowski, Pulich, and Van Baalen 1977).

Although the results of this investigation were similar to other studies cited above in that a gradational nature was defined for the Texas shelf benthos related to several environmental variables, differences between some of these variables in this and other studies was the key. As stated earlier, the Texas shelf differs from other shelf ecosystems because of the silty nature of its sediments. The infaunal-environmental relationships observed here suggest that these siltier sediments may be responsible for a difference in dominant taxa on the Texas shelf compared with those of other shelves such as the middle Atlantic.

Polychaetes were the dominant taxa observed in this study. The majority of their feeding strategies, according to comparisons with strategies of the fauna discussed by Fauchald and Jumars (1979), involved deposit feeding modes. These strategies are much more conducive to silty, unstable bottom habitats (Sanders 1960; Saila 1976). In contrast, the dominant fauna observed on the middle Atlantic shelf were amphipods (Boesch 1979). This shelf is characterized by sandier sediments than the Texas shelf. Amphipods derive their nutrition primarily by suspension feeding, which according to Sanders (1960) and Levinton (1972) is a more appropriate feeding strategy for sandier, more stable sediments.

It was concluded that the subtropical Texas shelf showed infaunal patterns consistent with other shelf ecosystems in terms of environmental gradation (Day, Field, and Montgomery 1971) and shallow water variability as found in temperate marine systems (Sanders 1968). The Texas shelf differed, however, from at least one other shelf extensively studied (Boesch 1979) in that a different taxa dominated the infauna, and this difference was possibly related to the sediment structure differences of the mid-shelf habitat between the two areas.

EPIFAUNA   Northern Gulf of Mexico epifauna is considered by many investigators an extension of the Carolinian province with faunal divisions at the Mexican border and just east of the Mississippi delta (Hedgpeth 1953). The STOCS study area falls within the Texas to Mississippi delta region but by virtue of the southernmost stations is influence by Caribbean fauna. Distribution of any species is based on a complex of environmental factors. Temperature and salinity control the range of benthic species, but within that range, more subtle factors determine faunal distribution. Depth was the most apparent factor controlling epifaunal distribution in this study.

The results of cluster analyses, which were used to define community changes for epifauna on the Texas shelf, were relatively similar for both 1976 and 1977. The analyses divided the shelf into two major regions based on depth or distance from shore or both (Figure 42). All stations within a 10- to 45-m depth (plus 2/II at 49 m) and located less than 48 km offshore were grouped together (A). Stations with depths greater than 45 m and located at least 48 km offshore formed the other major group (B). The two regions varied in other physical variables. Bottom water temperatures (10° C to 29° C) and salinity (30‰ to 37‰) varied widely throughout the year in Group A. Group B stations were characterized by a more stable temperature (15° C to 25° C) and salinity (35‰ to 37‰) regimen. There was considerable overlap of sediment types between the two regions; however, the sandiest sediments were found at shallow stations near shore, and the highest clay content was in sediments from the deep offshore stations.

Subdivisions from cluster analysis divided the study area into six groups of stations (Figure 42). These minor divisions generally corresponded to shallow (10 to 15 m in depth), shallow-intermediate (22 to 45 m), deep-intermediate (47 to 100 m), and deep (106 to 134 m) stations. Stations that clustered together two out of three seasons were considered to be a group. Seasonal changes in abundance and the mobility of epifaunal organisms precluded clean distinctions of station groups consistent throughout the year.

Station groups were also defined by the species found there. Clustering by species or inverse analysis resulted in eight species groups (Figure 43). The first three groups of species were collected only at stations greater than 45 m in depth; species Group IV was taken most consistently at the same stations. Group V species were collected at

[108] Benthic Biota

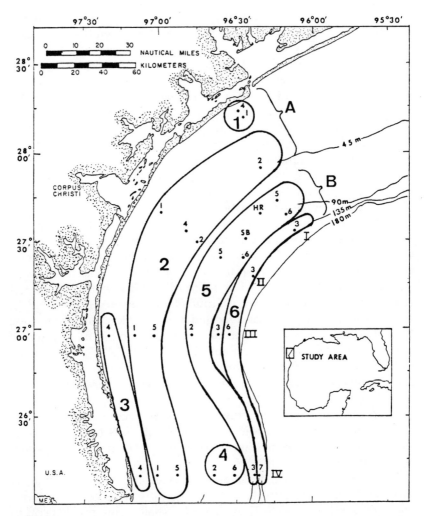

Figure 42. Location of station groups from cluster analysis of seasonal epifaunal data.

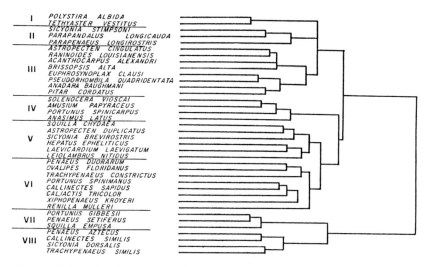

Figure 43. Species dendrogram from inverse analysis of seasonal epifaunal data. Roman numerals refer to species groups.

intermediate depth stations but not at shallow or deep stations. Species in Groups VI and VII were most often collected at stations less than 45 m in depth. Group VIII species were collected at all but the deepest stations. Although a species group was relatively constant to a station group, most individual species responded in a unique way to the physical environment common to the stations.

For example, many of the species most characteristic of the shallow shelf are motile decapods found in inlets, bays, and shoal areas in summer and early fall. Copeland (1965) collected large numbers of *Trachypenaeus similis* and *Squilla empusa* in Aransas Pass Inlet in late summer and early fall. Large numbers of *Penaeus setiferus* are found in the bays in fall and support a sizable bay fishery. Seasonal changes in population may be related to the annual temperature (14° C to 29° C in 1976) and salinity (31‰ to 36‰ in 1976) extremes at inner-shelf stations. In contrast, large numbers of species with low abundance characterize the outer-shelf assemblage. The high equitability and species richness of this area reflect the relatively stable environment conditions characteristic of the area.

Corresponding to the Texas shelf's general community structure differences remaining relatively consistent over two years of study were specific community variables such as number of species, den-

[110] Benthic Biota

sity, diversity, and equitability, which showed similar trends across the shelf between years (Figures 44 through 46). The number of epifaunal species collected per station presented no consistent general pattern. The numbers collected were much smaller than those of the infaunal species. Along Transect I, epifaunal species numbers were fairly evenly distributed during each collection period. Differences between seasons were apparent. The winter sampling showed fewer species collected on this transect. The number of species collected at 5/I was somewhat depressed at all collection times. Transect II showed a varying pattern of species abundance spatially and temporally. The winter collection had a peak species abundance at 6/II, suggesting an increase with depth. There were species abundance peaks at Station 2/II at both spring and fall collections so that the abundance of species was greatest at mid depth and decreased shoreward and offshore. On

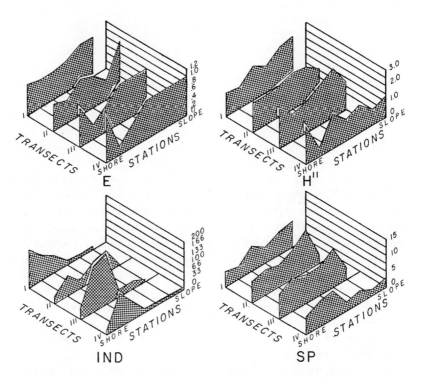

Figure 44. Shannon diversity values, H''; equitability, E; and number of species (SP) and number of individuals (IND) for winter 1977 epifaunal data.

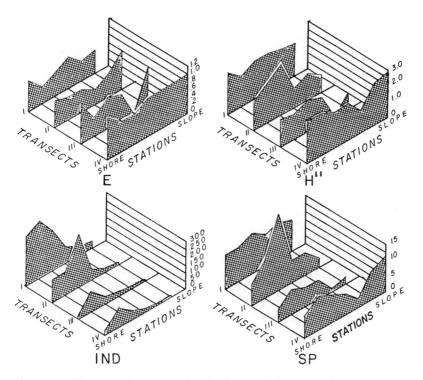

Figure 45. Shannon diversity values, H''; equitability, E; and number of species (SP) and number of individuals (IND) for spring 1977 epifaunal data.

Transect III, there was a slight winter increase in species richness with depth. Spring collections at the deepest stations (3 and 6) were extremely depressed. Minor peaks in species abundance occurred at Stations 1/III and 2/III in the fall. Transect IV epifaunal species richness varied widely with season. The winter collection showed a strong decrease in species richness with depth. In spring, numbers of species were somewhat evenly distributed along the transect, and Station 6/IV numbers were somewhat depressed. The fall collection exhibited a strong positive correlation of species richness to water depth.

The number of individual epifaunal organisms collected at each station generally peaked at mid depth or shallow-intermediate depths and decreased shoreward and offshore (Figures 44 through 46). Transect I epifaunal density distributions followed the same general pattern but with maximal numbers of individuals at inshore sites, more

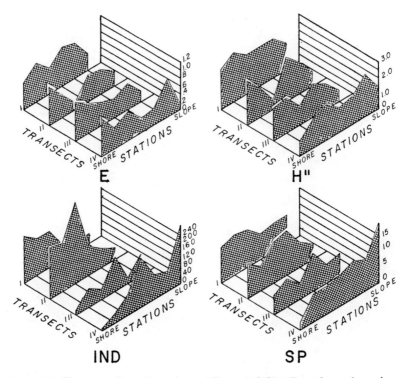

Figure 46. Shannon diversity values, H''; equitability, E; and number of species (SP) and number of individuals (IND) for fall 1977 epifaunal data.

so than other transects. Transect II demonstrated a highly varied pattern of density. Winter collections were small and similar in numbers of individuals across the shelf. A major peak in density occurred in the spring collection at Station 2/II, with numbers of individuals decreasing shoreward and offshore. The fall collection on this transect displayed the densest epifaunal communities observed for the entire year, particularly at the four shallowest stations. The number of epifaunal organisms collected on Transect II varied widely through the year. Dense populations on the inner half of the study area occurred in the winter collection. Densities along this transect were low during the spring collection period. Fall collections indicated peaks of abundances at Stations 1, 2, and 6. Epifaunal density patterns on Transect IV were similar to those of Transect III, the winter onshore collections slightly diminishing and the fall collections generally increasing.

Epifaunal diversity (H'') was generally lower than that exhibited by infaunal collections. No general pattern of diversity across all transects was observed. Relatively high densities were consistent across the shelf for Transect I, although there existed some tendency to increase with depth in the winter, a major peak at Station 2/II in the spring, and a major decline at Station 3/II in the fall. Transect III exhibited a fairly high winter diversity but a minimum at Station 1/III. Increases in this variable with depth were observed, except at the deepest site, where there was a decrease in diversity. A major decrease in diversity was observed at 3/III in the spring collection. Diversity on Transect III in the fall was fairly uniform with decreases at Stations 5 and 6. There were relatively low diversities on Transect IV in the winter, uniformly high values in spring with the exception of Station 3/IV, and a tendency of H'' values to increase with depth in the fall.

Epifaunal equitability values showed no pattern consistent to all transects. There was a trend toward greater equitability inshore and offshore with mid-depth areas being depressed. A smooth pattern in the winter of increasing equitability with depth was evident on Transect I. Spring and fall values were more diverse, with low equitability at Stations 1 and 6 in the spring and Stations 4, 2, and 3 in the fall. Transect II exhibited increased equitability at the deepest site (Station 3) in winter and spring that decreased sharply in fall concomitant with an increase in equitability at the nearshore stations. Transects III and IV were very similar in the winter, demonstrating high equitability at all except the shallow mid-depth stations. Equitability on Transect III showed peaks shallow and deep with variable levels between in the spring, while Transect IV values remained almost uniformly high. Fall collections on Transects III and IV indicated the trend toward greater equitability inshore and offshore and decreased values at mid depths.

As illustrated above, epifaunal community structure variables showed no general trends or spatial patterns. Variation in temporal and spatial abundances of dominant species in 1976 was due to recruitment of young age classes at shallow to shallow-intermediate stations, as well as to migration of the adult population, accompanied by reduction in abundance, to the deeper stations. The same pattern was observed in 1977, except there was a stronger tendency for the abundance to be concentrated at stations along Transects I and II.

Because epifaunal species differ in physical and biological needs and some are capable of moving considerable distances, analysis of individual species distributions may be the best method for interpreting the STOCS data. An important aspect of this study was the evaluation of the shelf ecosystem in terms of numbers and kinds of species. Further information that can be derived from the data includes an understanding of the distribution of species important to man (directly or indirectly) and the identification of species with narrow or critical tolerances to environmental change.

Demersal Fish

Patterns of distribution and abundance of outer continental shelf fishes off the Texas coast have been examined by a number of workers (e.g., Hildebrand 1954, Chittenden and Moore 1977), but ecological aspects of this ichthyofauna (particularly regarding factors that affect these patterns) remain poorly understood. Although distributions of certain species off Texas and in other parts of the Gulf of Mexico have been related in a broad manner to a few obvious factors such as depth, sediment, and temperature (Dawson 1964; Chittenden and McEachran 1976, Lewis and Yerger 1976), future planning demands a more detailed exposition of the relationship between fish populations and the ecological factors that affect them. The need for statistical evaluations of these fish-environment relationships has been specifically pointed out (Chittenden and Moore 1977).

The numerical classification of data for demersal fishes indicated that distinct station groups, aligned with depth, could be identified on the basis of ichthyofaunal composition. Although the exact delineations between station groups varied among time periods (day with night, season, year), the general separation of groups by depth appeared to be a consistent feature. Therefore, a grouping of stations by depth (Figure 47) was adopted for use in subsequent analyses.

Other general conclusions that emerged from numerical classification analyses are (1) although zonation appeared primarily to be depth related, temperature and seasonal migration patterns apparently also influenced the species associations; (2) the shallow-shelf turbulent zone exhibited low species diversity throughout the year and especially high numbers of individuals in winter and spring; (3) the faunal associations near shore dissipated during the late summer or autumn when shallow water temperatures were highest; (4) mid- and deep-water associations were somewhat more stable throughout the year, mid-shelf groups of species having the highest diversity; (5) north-south gradients were minimal except during autumn when weak species associations developed to show that the two northern transects were slightly different from the two southern transects; and

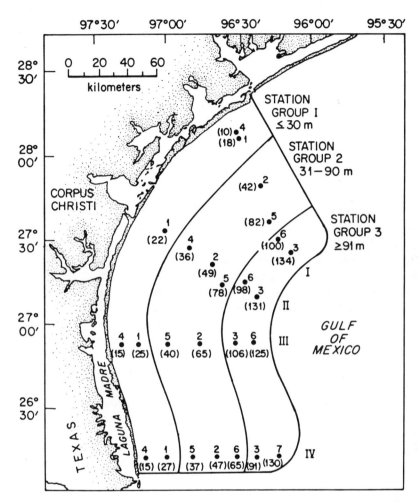

Figure 47. Station groupings for demersal fish according to cluster analysis. Numbers in parentheses are depths in meters.

(6) there was evidence of considerable species "shuffling" during the year in all faunal zones, which suggested that species-dominated communities, like those described by Petersen (1913 and 1918), did not persist in the shelf areas that were studied.

Over 160 fish species were captured during the three years of sampling, but only 57 species were captured in excess of 100 individuals and 22 species in excess of 1000. The most common species are listed in Table 16, and their frequencies of occurrence among the ten most abundant species for each season and station group (defined by Figure 47) are given in Tables 17 and 18. Most of the common species were dominant elements (i.e., among the top ten species) of the ichthyofauna in only one or two station groups (e.g., *Syacium gunteri* and *Diplectrum bivittatum* in Station Groups 1 and 2; *Serranus atrobranchus* in Station Groups 2 and 3), reflecting the spatial (according to depth) differences in the fish assemblages found over the study area. The notable exception was *Trachurus lathami*, which was dominant at roughly equal frequencies in all three station groups. Seasonal changes in the extent to which species dominated the ichthyofauna also occurred (e.g., *Syacium gunteri* was predominant mainly in winter and fall), although a number of species showed no variation (e.g., *Stenotomus caprinus* and *Serranus atrobranchus*; see Table 17).

A greater number of species were caught in night trawls than in day trawls during the seasonal sampling cruises. Part of this difference was attributed to the greater sampling effort expended during the night cruises. However, the difference between the number of night trawls taken and the number of day trawls taken during the seasonal cruises was rather small, and it appeared reasonable to conclude that some biological reason existed for the observation that night trawls yielded greater numbers of species than did day trawls during both 1976 and 1977.

In terms of contrasting day with night collections of demersal fishes, variables such as biomass and number of species and individuals as well as measures of diversity differed throughout the year. Fish taken predominantly during day collections were commonly schooling species, while predominantly nocturnal species were solitary in nature. Numbers of species were low in fall and high in the spring.

General catch statistics illustrated that highest densities of demersal fish occurred during the day in spring and during the night in fall. The lowest catches occurred in winter for both day and night. The lowest biomasses were taken during the winter for both day and night. Highest seasonal biomasses were observed during the night in fall. Spring and fall daytime collections yielded much higher biomasses than did collections in winter.

Table 16. Total Abundance and Occurrences (Number of Trawls in which Taken) of the Most Abundant Fishes Captured During the Sampling Program, 1975–1977

Species	No. of Individuals	No. of Occurrences
Trachurus lathami	8,612	243
Serranus atrobranchus	8,406	365
Micropogon undulatus	7,767	140
Peprilus burti	6,656	169
Cynoscion nothus	5,952	123
Syacium gunteri	4,465	263
Stenotomus caprinus	3,905	327
Pristipomoides aquilonaris	3,534	312
Prionotus paralatus	2,608	235
Polydactylus octonemus	2,392	65
Saurida brasiliensis	2,162	194
Anchoa hepsetus	1,987	59
Chloroscombrus chrysurus	1,945	65
Sphoeroides parvus	1,724	163
Upeneus parvus	1,724	217
Centropristis philadelphica	1,705	297
Prionotus stearnsi	1,635	187
Cynoscion arenarius	1,431	130
Prionotus rubio	1,429	217
Trichopsetta ventralis	1,390	193
Synodus foetens	1,186	308
Diplectrum bivittatum	1,072	133
Porichthys porosissimus	957	189
Pontinus longispinis	548	73
Synodus poeyi	512	112
Bollmannia communis	507	112
Lepophidium graellsi	455	149

No obvious trends correlated biomass to depth. The relationships between abundances of selected fish species and some physical variables were examined by plotting abundances of these species against values of a particular variable. From this exercise it was apparent that within a given species different sizes of fish may respond differently to some environmental variables. The implication is that further studies should consider the possible effects of individual size on the relations of the fish to environmental conditions.

The diversity of demersal fishes was low at shallow stations but increased with depth to about 85 m. As with the epifauna, seasonal changes in fish populations appeared to be related to depth, temperature, and movements into and out of the estuaries.

Table 17. Frequency of Occurrence of Common Fishes Among the Ten Most Abundant Species During Each Season[a]

Species	No. of Occurrences Among Top Ten Species[b]		
	Winter	Spring	Fall
Anchoa hepsetus	2	3	1
Cynoscion nothus	3	3	3
Micropogon undulatus	3	4	6
Peprilus burti	3	6	3
Syacium gunteri	8	3	7
Cynoscion arenarius	4	3	1
Sphoeroides parvus	4	3	4
Trachurus lathami	2	8	6
Polydactylus octonemus	—	4	6
Chloroscombrus chrysurus	—	3	3
Upeneus parvus	4	7	4
Stenotomus caprinus	8	8	9
Diplectrum bivittatum	2	1	6
Saurida brasiliensis	3	2	2
Serranus atrobranchus	8	8	8
Synodus foetens	2	3	1
Prionotus stearnsi	2	3	5
Pristipomoides aquilonaris	6	7	6
Prionotus paralatus	4	5	5
Trichopsetta ventralis	5	3	4
Halieutichthys aculeatus	2	1	3
Pontinus longispinis	1	2	4
Prionotus rubio	4	1	3
Centropristis philadelphica	3	3	3

[a] Data from Wohlschlag et al. 1977 and Wohlschlag et al. 1979.
[b] Each occurrence corresponded to a single sampling period (e.g., winter–day–1977). Twelve occurrences per station were possible. Sampling periods included winter, spring, fall/day, night/1976, 1977.

Discriminant analysis, using the defined demersal fish station groups in Figure 47, was applied to data on STOCS bottom water and sediment physical variables to investigate relationships between the fish communities and their habitat. Physical variables used were bottom water temperature and salinity, sediment mean grain size, standard deviation and skewness of the sediment grain size distribution, and percentage of silt composing the sediment.

Table 18. Frequency of Occurrence of Common Fishes Among the Ten Most Abundant Species in Each Defined Station Group[a]

Species	No. of Occurrences Among Top Ten Species[b]		
	Station Group 1	Station Group 2	Station Group 3
Anchoa hepsetus	6	—	—
Cynoscion nothus	8	1	—
Micropogon undulatus	11	2	—
Peprilus burti	7	4	1
Syacium gunteri	9	8	1
Cynoscion arenarius	7	1	—
Sphoeroides parvus	7	4	—
Trachurus lathami	5	6	5
Polydactylus octonemus	7	3	—
Chloroscombrus chrysurus	5	1	—
Upeneus parvus	2	5	8
Stenotomus caprinus	2	12	11
Diplectrum bivittatum	4	5	—
Saurida brasiliensis	—	5	2
Serranus atrobranchus	—	12	12
Synodus foetens	—	5	1
Prionotus stearnsi	—	8	2
Pristipomoides aquilonaris	—	7	12
Prionotus paralatus	—	2	12
Trichopsetta ventralis	—	2	10
Halieutichthys aculeatus	—	1	5
Pontinus longispinis	—	—	7
Prionotus rubio	4	2	2
Centropristis philadelphica	2	7	—

[a] Data from Wohlschlag et al. 1977 and Wohlschlag et al. 1979.
[b] Each occurrence corresponded to a single sampling series (e.g., station group 1–day–1977). Twelve occurrences per season were possible. Sampling series included Station Groups 1, 2, 3/day, night/1976, 1977.

The analysis yielded two discriminant functions. Values for the standardized coefficients indicated that mean grain size, salinity, and percentage silt were the most important variables on the first discriminant function. These three variables served most to discriminate between demersal fish station groupings on the shelf and the first discriminant function. Mean grain size, skewness, and standard de-

[120] Benthic Biota

viation of the grain size distribution served most to separate demersal fish station groups on the second discriminant function.

The discriminant scores of the data cases for the three fish station groupings are plotted against the first and second discriminant functions in Figure 48. Each of these groups was found to be significantly different from one another. Differences by pairs between station group centroids were tested for significance using F values based on the Mahalonobis distance between groups. Each station group showed significant differences from the two other groups ($P < 0.001$).

Discriminant analysis using fish abundances (from all sampling episodes over three years) as discriminating variables yielded two discriminant functions, with the first approximately twice as important as the second in separating station groups. Standardized co-

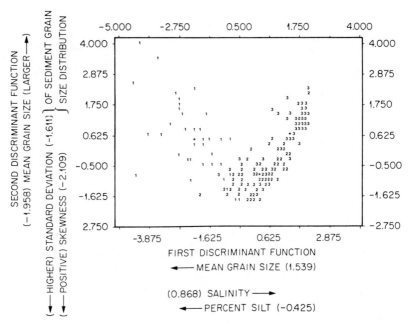

Figure 48. Discriminant space based on analysis of physical variables. Each number represents a sampling episode (sampling station for a given time period; e.g., Station 1/I, day–winter) with value equal to station group identification. Symbol (+) denotes a group centroid. Values in parentheses by discriminating variables are standardized discriminant function coefficients (weights).

efficients showed the following species to contribute most to the first discriminant function: *Cynoscion nothus, Pristipomoides aquilonaris, Sphoeroides parvus, Syacium gunteri, Trichopsetta ventralis*, and to a lesser degree, *Chloroscombrus chrysurus, Micropogon undulatus, Peprilus burti*, and *Prionotus paralatus*. *Bollmannia communis, Syacium gunteri, Synodus foetens, Synodus poeyi*, and *Trichopsetta ventralis* were the most important species for the second discriminant function.

Centroids of the station groups and discriminant scores for data cases are plotted in Figure 49. Statistical comparisons by pairs of the group centroids (using F values based on Mahalonobis distances be-

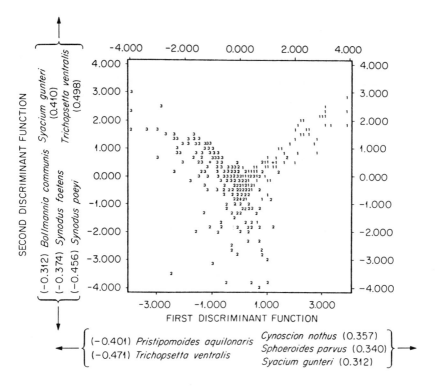

Figure 49. Discriminant space based on analysis of fish abundance. Each number represents a sampling episode (sampling station for a given time period; e.g., Station 1/I, day–winter) with value equal to station group identification. Symbol (+) denotes a group centroid. Values in parentheses by discriminating variables are standardized discriminant function coefficients (weights).

tween groups) revealed significant differences between all groups ($P < 0.001$).

Although the discriminant analysis using fish abundance data was aimed toward obtaining descriptions of the defined station groups and the common fishes, it also provided a test of the growth of the groupings (as did the analysis using physical variables). The statistically significant differences between the groups ($P < 0.001$) and the moderately high proportion (0.758) of the data cases correctly classified in the discriminant analysis indicated that the defined demersal fish station groups (Figure 47) could be satisfactorily differentiated on the basis of abundances of common fishes while being explained in terms of their relation to environmental factors of the benthic habitat.

The work outlined above on demersal fishes serves to identify major components of the outer continental shelf benthic ichthyofauna and describes the more obvious spatial and temporal patterns in abundance of species. The characterization of major depth zones by discriminant analysis using common fishes as well as by straightforward descriptions of the fauna should be particularly useful for the assessment of man-induced impacts on this environment.

Benthic Biota Body Burdens

HYDROCARBONS The majority of the studies regarding petroleum pollution have centered on the immediate and long-term effects of catastrophic events such as oil spills. This emphasis is partly due to the identifiably apparent impact of large amounts of oil on an area and partly due to the relative ease of identifying and quantifying some petroleum compounds in spill situations. The effects of low-level and chronic input of petroleum have been less intensively studied, and information on background levels of hydrocarbons in unpolluted environments is scarce. Although identifying and quantifying trace quantities of petroleum hydrocarbons have been major deterrents to low-level studies, methods are rapidly being developed for hydrocarbon trace analyses.

One of the major problems associated with quantifying trace levels of petroleum in the environment is differentiating petrolic compounds from biogenic hydrocarbons. This differentiation is complicated by the effects of weathering or environmental degradation on the hydrocarbon composition of petroleums. Unlike the case of an oil spill, in which a single source of petroleum generally provides a very characteristic hydrocarbon pattern, trace levels of petroleum may be from a number of sources, such as petroleum production or shipping and waste disposal, which further complicates hydrocarbon patterns and thus detection and quantification.

The use of a number of variables has been suggested to aid the analyst in distinguishing sources of hydrocarbons in environmental samples. One of these is the measurement of ratios of concentrations of individual hydrocarbons, such as the ratio of n-heptadecane (C_{17}): pristane and of pristane:phytane (Ehrhardt and Blumer 1972). These ratios have been suggested as aids in detecting a single source of petroleum contamination because they are generally characteristic of an oil.

One study of importance (Gilfillan et al. 1977) indicated that concentrations of hydrocarbons in clams collected from various areas did not correlate to those in the sediments. Concentrations of hydrocarbons in clams were found to range from 8.5 to 11 µg/g body weight, while concentrations in sediments ranged from 9 to 228 µg/g. Benthic organisms collected from unpolluted deep sea areas had hydrocarbon distributions quite different from those distributions found in surrounding sediments (Teal 1976). These reports indicate that the effect of sediment-absorbed hydrocarbons on the hydrocarbons of benthic epifauna is quite difficult to predict. It is probable that hydrocarbons in water, including those from sediment desorption in interstitial water, have a greater effect on the hydrocarbon content of benthic organisms than those absorbed directly from sediment. Uptake from food is also probably a more important source of hydrocarbons than uptake from sediment.

Information on mechanisms of hydrocarbon transport in the marine environment, such as food chain transfer and uptake from water and sediment, is important for assessing the probable effects of petroleum hydrocarbons. Few studies of hydrocarbon distributions and transport, particularly in benthic organisms and the benthic environment, have been reported.

Approximately 400 faunal samples from the benthic habitat, representing 48 species, were analyzed for high molecular-weight hydrocarbons. The mean and standard deviations of selected variables from the analyses of the six most frequently occurring species are given in Table 19. Overall, total hydrocarbon concentrations ranged from less than 0.01 µg/g (ppm) to 54.47 µg/g dry weight with the majority of samples containing less than 1 µg/g. Pentadecane (C_{15}) and heptadecane (C_{17}) were the dominant n-alkanes, frequently constituting 70% or more of the alkanes. Pristane was found in almost all samples at relatively high levels. Phytane was found in approximately 20% of the samples, generally at less than 0.05 µg/g. The

Table 19. Means and Standard Deviations for Selected Variables from Heavy Molecular-Weight Hydrocarbon Analyses of Macroepifauna and Macronekton

Species (N[a])	Total Alkanes (μg/g) (N)	Sum of Alkanes (%)		
		$C_{14}-C_{18}$ (N)	$C_{19}-C_{24}$ (N)	$C_{25}-C_{32}$ (N)
Loligo spp. (45)	1.89 ± 3.32 (45)	60.9 ± 34.0 (42)	16.5 ± 18.6 (42)	22.7 ± 28.5 (42)
Penaeus aztecus (48)	0.14 ± 0.28 (48)	39.7 ± 7.2 (34)	9.1 ± 12.7 (34)	51.1 ± 39.1 (34)
Pristipomoides aquilonaris (38)	2.98 ± 4.28 (38)	82.7 ± 26.3 (38)	7.1 ± 10.3 (38)	10.2 ± 3.2 (38)
Serranus atrobranchus (27)	0.19 ± 0.20 (27)	69.4 ± 32.7 (21)	9.7 ± 11.8 (21)	20.9 ± 28.8 (21)
Stenotomus caprinus (27)	1.02 ± 1.83 (27)	55.2 ± 31.4 (26)	13.3 ± 18.1 (26)	31.5 ± 26.3 (26)
Trachurus lathami (25)	8.58 ± 12.00 (25)	69.0 ± 36.3 (25)	13.5 ± 17.7 (25)	17.5 ± 23.7 (25)

[a] N, number analyzed.
[b] CPI, carbon preference index.

pristane : phytane, pristane : heptadecane, and phytane : octadecane ratios ranged widely and did not appear to be indicative of a common source of petroleum in the study area. The carbon preference index (CPI) ratio, illustrating odd-carbon dominance especially for the CPI_{14-20} and CPI_{20-32} ratios, also was not indicative of petroleum contamination. Squalene was frequently the only compound detected in the aromatic fraction. Aromatic compounds were rarely detected and were usually at 0.005 μg/g or lower concentrations. The distribution of aromatics was not suggestive of petroleum origins. Unresolved complex mixture peaks were also rarely detected in the gas chromographs and were very low when present. The distribution of phytane in the samples appeared to yield a spatial trend. Phytane was found most frequently at Stations 1 and 2 of Transects III and IV and more

Pristane Phytane (N)	Pristane C_{17} (N)	Phytane C_{18} (N)	CPI^b_{14-20} (N)	CPI^b_{20-32} (N)
166.5 ± 172.9 (2)	9.3 ± 13.4 (34)	2.7 ± 3.2 (2)	18.6 ± 10.4 (23)	3.7 ± 4.2 (27)
44.0 ± 58.0 (2)	2.0 ± 1.7 (17)	0.2 ± 0.1 (2)	1.6 ± 0.6 (5)	6.8 ± 10.1 (22)
46.7 ± 20.5 (5)	2.6 ± 2.4 (34)	0.6 ± 0.1 (5)	16.0 ± 7.2 (28)	6.7 ± 18.3 (18)
13.0 ± 5.6 (2)	3.6 ± 6.0 (18)	1.0 ± 0.7 (2)	6.2 ± 3.5 (8)	1.4 ± 0.8 (8)
32.8 ± 25.3 (6)	5.9 ± 3.9 (25)	0.9 ± 0.2 (5)	5.6 ± 3.3 (16)	10.1 ± 21.8 (17)
132.0 ± 65.9 (10)	33.4 ± 37.0 (18)	2.1 ± 0.8 (10)	19.3 ± 6.9 (15)	6.2 ± 8.6 (17)

frequently at Stations 1 and 2 of Transects I and II than at Station 3 of all transects.

Analyses of variance, testing for spatial and temporal differences, indicated three correlations for brown shrimp (*Penaeus aztecus*) that appeared to be good indicators of seasonal changes in hydrocarbon distributions. As can be seen in Figure 50, the hydrocarbon distribution changes with season, causing significant changes in the low and high sums of hydrocarbons ($P = 0.02$) and the CPI_2 ($P = 0.01$).

The dominant hydrocarbons were pristane, pentadecane, and heptadecane. These hydrocarbons probably reflect dietary sources since pristane is the major hydrocarbon in zooplankton (Blumer, Mullin, and Thomas 1964), and pentadecane and heptadecane are the major hydrocarbons in unpolluted algae (Clark and Blumer 1967). The over-

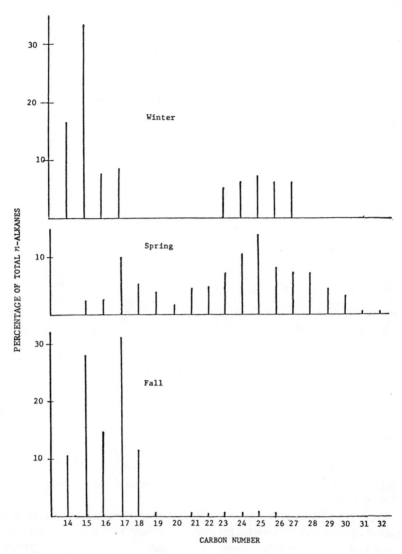

Figure 50. Percentage distribution of *n*-alkanes in *Penaeus aztecus* (brown shrimp).

all concentration of hydrocarbons in the samples was generally quite low (less than 1 µg/g dry weight in many samples), and the hydrocarbon distributions found were not suggestive of petroleum. The CPI ratios showed the high odd-carbon dominance characteristic of biogenic hydrocarbons (Clark 1974; Clark and Finley 1973; Cooper and Bray 1963), although shrimp tended to have CPI_{14-20} values close to 1.

Phytane was the only potential indicator of petroleum (Farrington et al. 1972; Blumer and Snyder 1965) found with any frequency in the samples. It was found most often in samples from Stations 1 and 2 of all transects. This finding may indicate some petroleum contamination from onshore or shipping activities or may reflect species variation and mobility, since the species collected at Stations 1 and 2 were generally different from those at Station 3 of all transects.

The distribution of aromatics, when present, was not suggestive of petroleum sources. Thus, petroleum contamination of the benthic organisms of the study was not significant during the study period, and the data obtained should provide an excellent data base for future studies of petroleum pollution. Researchers have concentrated their data synthesis efforts toward maximizing the data's utility for characterization purposes.

The significant results for brown shrimp are indicative of a change in hydrocarbon distribution that occurs in shrimp in spring, possibly due to spawning activities or to dietary effects (Figure 50). The hydrocarbon levels in shrimp were also lower in winter (0.04 µg/g) and fall (0.06 µg/g) than in spring (0.33 µg/g), although the differences were not significantly different ($P = 0.05$).

From the results of this study, shrimp appear to be excellent organisms for monitoring the presence of petroleum hydrocarbons. Shrimp demonstrate significant changes in hydrocarbon distribution with season, but these changes are relatively consistent and quantifiable. The low levels of hydrocarbons present in shrimp may also simplify the detection of pollutant hydrocarbons. A rig-monitoring sample obtained in winter and after drilling had 0.6 µg/g total hydrocarbons compared with 0.04 µg/g found for the winter samples in this study. This sample also had very low CPIs ($CPI_{14-20} = 1.1$, $CPI_{20-32} = 0.6$) and a distribution of hydrocarbons suggestive of petroleum, especially when compared with the patterns found for shrimp in this study, as shown in Figure 51. In contrast, shrimp from an oil-producing area of the Gulf had higher hydrocarbon levels (0.53 to 2.45 µg/g) (Middleditch and Basile 1978) than those found in this study.

Of the approximately 140 macronekton analyses performed, 120 were for two species, the red and vermilion snappers (*Lutjanus camp-*

[128] Benthic Biota

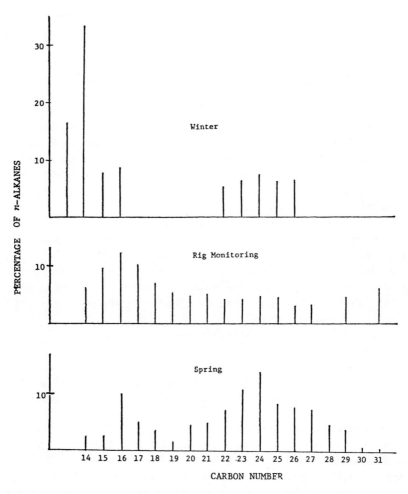

Figure 51. Comparison of rig monitoring and seasonal hydrocarbon distribution in *Penaeus aztecus* (brown shrimp).

echanus and *Rhomboplites aurorubens*). Approximately 20 samples were obtained for each species over two years of this study. Each sample yielded three tissue analyses: muscle, liver, and gill in 1976, and those and gonad in 1977, which were each analyzed separately. Table 20 summarizes the ranges and means of total hydrocarbon concentrations found in macronekton. The means of several of the variables measured in muscle and liver are shown in Table 21. The alkanes, *n*-pentadecane (C_{15}) and *n*-heptadecane (C_{17}), and pristane were the major aliphatic hydrocarbons in all samples. In red snapper muscle, the C_{15} plus the C_{17} *n*-alkanes totaled 23% to 100% of the *n*-alkanes; the total was less than 75% in only 3 of the 20 samples. One of these samples had the C_{27} *n*-alkanes as the major *n*-alkane, but the other two had a wide range of hydrocarbons. In vermilion snapper muscle, C_{15} plus C_{17} ranged from 42% to 100% of the *n*-alkanes. The C_{19} or C_{23} alkanes had relatively high concentrations in the two samples with the lowest C_{15} plus C_{17} concentrations.

A wider range of hydrocarbons, as well as higher concentrations of the C_{18}–C_{30} *n*-alkanes, appeared to be present in the spring samples rather than the fall or winter samples, but the differences were not statistically significant ($P < 0.05$). The liver, gill, and gonad samples also had pentadecane, heptadecane, and pristane as the major hydrocarbons with nonsignificant seasonal changes similar to those found in muscle. Phytane was found in 10% of the muscle samples, in more than 50% of the liver samples, and in all of the gonad samples, generally at low concentrations. Small, generally unquantifiable, unresolved complexes were detected in many of the chromatograms. Aromatic compounds were rarely detected and, when found, were generally at the limits of detection (0.005 µg/g).

The high molecular-weight hydrocarbon analysis of macroepifauna and macronekton samples from the STOCS indicated little, if any, petroleum contamination of the study area. No significant spatial trends and few seasonal trends were present in the data, suggesting relative stability in the hydrocarbon pools of the organisms studied. Of the species studied, the brown shrimp *Penaeus aztecus* appears to be the best indicator for monitoring purposes.

TRACE METALS   Ten trace elements were analyzed in benthic biota including cadmium (Cd), chromium (Cr), copper (Cu), iron (Fe), nickel (Ni), lead (Pb), vanadium (V), zinc (Zn), aluminum (Al), and calcium (Ca). Nickel and V were selected because they are present in large concentrations in some oil and tars. Cadmium and Pb, two very toxic metals, are frequently observed to be above natural levels near indus-

Table 20. Ranges of Total Hydrocarbon Concentrations in Macronekton

	Concentration ($\mu g/g$ dry weight)			
Species	Muscle	Liver	Gill	Gonad
Red Snapper (*Lutjanus campechanus*)				
Range	0.03–7.4	1.1–43.8	0.1–20.0	1.7–55.1
Mean ± SD	0.7 ± 1.6	8.7 ± 9.6	4.7 ± 6.1	36.8 ± 31.8
Vermilion Snapper (*Rhomboplites aurorubens*)				
Range	0.02–4.3	0.6–35.8	0.0–30.6	2.3–25.3
Mean ± SD	1.4 ± 1.3	13.6 ± 9.8	7.0 ± 9.4	6.8 ± 7.2

Table 21. Means and Standard Deviations for Selected Variables from Heavy Molecular-Weight Hydrocarbon Analyses of Macronekton

	Species and Organ			
	Lutjanus campechanus		Rhomboplites aurorubens	
Variable	Muscle	Liver	Muscle	Liver
$\Sigma\, C_{14-18}$	87.9 ± 25.6	83.4 ± 19.5	91.2 ± 12.5	80.3 ± 22.9
$\Sigma\, C_{19-24}$	4.5 ± 10.2	6.6 ± 7.8	5.0 ± 8.3	8.6 ± 11.4
$\Sigma\, C_{25-32}$	7.6 ± 19.6	10.0 ± 15.1	3.8 ± 8.9	11.0 ± 13.2
Pristane Phytane	13.5 ± 4.8	30.0 ± 24.1	87.7 ± 38.5	81.4 ± 59.1
Pristane C_{17}	1.9 ± 1.5	2.5 ± 1.5	16.3 ± 29.0	10.7 ± 9.9
Phytane C_{18}	0.6 ± 0.2	0.9 ± 0.9	1.3 ± 0.8	0.7 ± 0.5
CPI^a_{14-20}	16.7 ± 9.7	12.7 ± 10.5	22.6 ± 14.3	25.4 ± 35.7
CPI^a_{20-32}	1.0 ± 0.1	2.0 ± 2.2	1.5 ± 0.4	2.9 ± 4.1

[a] CPI, carbon preference index.

trial centers. Copper and Zn are essential trace metals that can reach toxic levels as a result of man's activities. Iron is also an essential trace element in biological systems (Dulka and Risby 1976; Brooks 1977). Iron and Al, because of their abundance in the environment, are important in making geochemical comparisons of trace elements (Trefry and Presley 1976b). Finally, Ca is important in identifying potentially severe matrix interferences in our analytical procedures.

Table 22 summarizes trace element data for the four selected species of demersal fish in terms of transect. The levels of several trace metals (i.e., Cd, Cr, Ni, Pb, V) in fish muscle were at or below the detection limits of our analytical procedures. For these metals it was obviously not possible to distinguish any spatial or temporal trends. Still, even for elements present in detectable amounts, none of the species exhibited any significant geographical patterns in muscle tissue trace element levels. *Trachurus lathami* was the only species to show any significant seasonal trends in trace metal concentrations.

Aluminum levels in demersal fish exhibited the same seasonal pattern as those in zooplankton. Aluminum and Fe in *Trachurus* muscle were strongly correlated ($r^2 = 0.41$). Also *Trachurus* was the only demersal fish species collected predominantly at stations near shore. Almost 90% of the samples came from Stations 1 and 2. These facts suggest that the temporal trend in Al levels was a reflection of the more variable environment near shore that is characterized by sizable seasonal fluctuations in the amount of suspended aluminosilicate particulate matter. The other three species of fish had generally similar concentrations of Al, but no seasonal trends were observed. These species were collected predominantly from offshore stations (i.e., 80% of the samples from Station 3), which are characterized by lower concentrations of organic-rich suspended matter.

Trace element data for penaeid shrimp muscle are summarized in Table 23 in terms of station and transect. No significant spatial trends in the data were detected for either species. *Penaeus setiferus* was only collected from the inshore stations. *Penaeus aztecus*, however, was consistently collected from 10 of the 12 stations sampled during this three-year study. Flesh trace element concentrations were not significantly different between the two species (i.e., paired t statistic, $P <$ 0.05). No strong correlations were observed between these data and corresponding sediment trace metal or potential prey organism variables. Aluminum and Fe levels in *P. aztecus* muscle were strongly correlated ($r^2 = 0.72$), and both metals exhibited significant correlations ($r^2 = 0.36$) with certain sediment texture variables. These results suggest that shrimp were assimilating sediment-derived Al and Fe into their muscle tissue.

[132] Benthic Biota

Table 22. Average Concentrations of Trace Elements in Muscle of Demersal Fish[a]

Transect	Species	No. of Samples	Cd	Cr	Cu
			(95% confidence interval observed		
I	Pristipomoides aquilonaris	7	<0.05	<0.05	1.3 (0.70–3.0)
	Serranus atrobranchus	4	<0.02	<0.05	0.95 (0.50–1.6)
	Stenotomus caprinus	7	<0.06	<0.05	1.0 (0.70–1.3)
	Trachurus lathami	2	<0.04	<0.03	2.4
II	Pristipomoides aquilonaris	23	<0.03	<0.04	1.4 (0.70–1.9)
	Serranus atrobranchus	8	<0.03	<0.05	0.90 (0.60–2.0)
	Stenotomus caprinus	11	<0.05	<0.05	1.1 (0.70–1.7)
	Trachurus lathami	8	0.10 (0.01–0.25)	<0.05	2.3 (1.7–3.0)
III	Pristipomoides aquilonaris	7	<0.03	<0.04	1.0 (0.60–1.6)
	Serranus atrobranchus	9	<0.02	<0.04	1.3 (0.50–3.5)
	Stenotomus caprinus	8	<0.06	<0.05	0.90 (0.60–1.1)
	Trachurus lathami	9	0.12 (0.01–0.30)	<0.10	2.5 (1.7–3.5)
IV	Pristipomoides aquilonaris	12	<0.05	<0.07	1.1 (0.60–2.0)
	Serranus atrobranchus	5	<0.02	<0.07	0.80 (0.50–1.6)
	Stenotomus caprinus	7	<0.05	<0.05	1.1 (0.60–1.5)
	Trachurus lathami	9	0.04 (0.01–0.09)	<0.07	2.2 (0.50–3.0)

[a] Concentration in ppm dry weight.

Benthic Biota [133]

Fe	Ni	Pb	V	Zn	Al	Ca
around mean)			(95% confidence interval observed around mean)			
4.5	<0.07	<0.04	<0.10	13	25	700
(2.0–6.0)				(11–16)	(19–30)	(400–950)
3.5	<0.09	<0.03	<0.10	10	30	1,100
(2.0–5.0)				(6.0–12)	(24–32)	(900–1,300)
5.5	<0.10	<0.08	<0.20	14	20	700
(4.0–6.0)				(11–17)	(15–25)	(400–1,200)
9.5	<0.08	<0.05	<0.15	24	19	800
4.0	<0.08	<0.04	<0.30	8.5	30	700
(1.0–7.0)				(1.0–17)	(16–45)	(350–850)
3.0	<0.08	<0.05	<0.40	10	30	1,900
(1.0–5.0)				(6.0–14)	(25–40)	(700–4,500)
5.0	<0.08	<0.06	<0.10	13	25	700
(2.0–8.0)				(6.0–25)	(12–55)	(400–1,400
15	<0.10	<0.06	<0.10	24	30	750
(8.0–20)				(12–35)	(15–40)	(550–1,000)
4.0	<0.07	<0.05	<0.10	10	22	500
2.0–7.0)				(2.0–16)	(16–30)	(300–600)
3.0	<0.09	<0.05	<0.10	10	20	1,300
2.0–4.0)				(2.0–17)	(14–30)	(750–2,000)
4.5	<0.08	<0.06	<0.10	13	13	600
4.0–5.0)				(10–18)	(10–20)	(350–850)
15	<0.10	<0.10	<0.15	24	25	1,000
7.0–25)				(15–40)	(10–50)	(300–2,500)
4.0	<0.08	<0.07	<0.10	14	18	600
1.8–6.0)				(10–20)	(15–25)	(300–1,100)
5.5	<0.10	<0.05	<0.15	12	17	1,800
4.0–6.0)				(11–16)	(14–30)	(750–3,500)
4.0	<0.08	<0.04	<0.10	12	23	1,200
3.0–6.0)				(6.0–15)	(20–25)	(750–1,600)
15	<0.09	<0.09	<0.10	19	17	800
6.0–25)				(13–25)	(12–30)	(550–1,200)

Table 23. Average Concentrations of Trace Metals in Flesh of Penaeid Shrimp[a]

Transect	Station	Species	No. of Samples	Cd	Cr	Cu
				(95% confidence interval)		
I	1	Penaeus aztecus	3	0.13 (0.10–0.20)	<0.05	25 (20–30)
		Penaeus setiferus	3	0.05 (0.01–0.10)	<0.05	21 (19–22)
	2	Penaeus aztecus	7	0.08 (0.01–0.20)	<0.05	25 (20–35)
	3	Penaeus aztecus	2	0.15 (0.13–0.17)	—	25 (20–30)
II	1	Penaeus aztecus	5	0.08 (0.02–0.12)	<0.05	24 (19–30)
		Penaeus setiferus	8	0.05 (0.01–0.12)	<0.05	24 (19–30)
	2	Penaeus aztecus	6	0.11 (0.02–0.25)	<0.05	25 (20–30)
III	1	Penaeus aztecus	4	0.07 (0.01–0.11)	<0.05	25 (25–30)
	2	Penaeus aztecus	5	0.10 (0.01–0.25)	<0.05	25 (18–35)
	3	Penaeus aztecus	4	0.18 (0.04–0.35)	<0.05	24 (18–35)
IV	1	Penaeus aztecus	4	0.06 (0.01–0.16)	<0.05	24 (20–28)
		Penaeus setiferus	1	0.01	<0.05	17
	2	Penaeus aztecus	6	0.08 (0.01–0.13)	<0.10	24 (18–30)
	3	Penaeus aztecus	5	0.18 (0.11–0.25)	<0.10	25 (20–30)

[a] Concentration in ppm dry weight.

Fe	Ni	Pb	V	Zn	Al	Ca
served around mean)			(95% confidence interval observed around mean)			
3.0	<0.10	<0.05	<0.07	45 (20–60)	18	1,200
3.5 2.0–5.0)	<0.10	<0.10	<0.05	50 (45–60)	20 (17–25)	950 (750–1,100)
4.5 1.0–10)	<0.10	<0.07	<0.20	50 (40–55)	20 (8.0–30)	1,100 (750–1,900)
—	—	—	—	50 (40–65)	—	—
3.5 3.0–4.0)	<0.15	<0.15	<0.30	55 (50–65)	36	2,500
2.5 .50–5.0)	<0.10	<0.10	<0.05	60 (50–70)	25 (15–35)	1,500 (450–2,500)
6.5 4.0–12)	<0.10	<0.10	<0.10	50 (40–60)	24 (14–34)	950 (800–1,100)
4.5 3.0–6.0)	<0.10	<0.05	<0.06	60 (55–65)	13	1,100
2.5 2.0–3.0)	<0.08	<0.10	<0.10	60 (50–70)	20 (17–25)	1,100 (850–1,500)
2.0	<0.10	<0.08	<0.12	50 (40–55)	20	1,400
2.0 1.0–3.0)	<0.08	<0.03	<0.10	55 (45–70)	22 (16–25)	700 (550–900)
1.0	<0.10	<0.10	<0.05	30	60	400
13 3.0–30)	<0.30	<0.15	<0.40	50 (45–60)	45 (13–90)	2,000 (450–3,500)
4.0	<0.30	<0.06	<0.20	50 (45–60)	24 (18–30)	1,000

Zinc levels in *P. aztecus* did exhibit a significant seasonal effect with a fall maximum. The reason for this relationship is not clear. The trend was not related to differences in the size (age) of shrimp analyzed among seasons. The seasonal fluctuations could have been a result of environmental changes that reflected physiological changes in the shrimp. Although no strong correlations ($r^2 < 0.20$) were observed between Zn levels and corresponding temperature, salinity, or dissolved oxygen conditions at the sampling sites, these variables were strongly correlated ($r^2 > 0.32$) to Zn concentrations in the hepatopancreas of the same shrimp.

One of the most striking aspects of this organismal trace element data set is the general lack of any significant spatial trends. This situation may in part be a result of the generally small number of data cases for many species, which made the detection of actual differences difficult. This absence of geographical trends, however, could be the result of at least two other factors. First, all of the species discussed here are quite mobile. Although the extent of their movements is generally not well documented, it certainly could be significant. This mobility would tend to integrate trace metal exposures at many sites and dampen any differences between them. Second, geographical trends in trace metal levels within the STOCS resulting from man's activities are probably minimal. Any significant input of trace metals to the STOCS area is most likely to be from diffuse (atmospheric) sources. Due to the relatively small amount of industrialization in the adjacent coastal areas, this atmospheric input is probably quite low and generally similar for all parts of the STOCS region.

It is worth noting that all of the significant seasonal trends observed for demersal fauna appeared to be linked to the more changeable environment near shore. Only species that were collected consistently at stations near shore exhibited any significant seasonality in trace metal levels. Species collected only at offshore stations showed no seasonal trends. This observation suggests that in any future monitoring it should be easier to detect changes in the levels of bioavailable trace elements at offshore stations than at ones near shore.

6
ECOSYSTEM CHARACTERISTICS

by R. W. Flint and N. N. Rabalais

General Trends

The three-year multidisciplinary study of the south Texas outer continental shelf resulted in the development of an extensive data base depicting the physical, chemical, and biological characteristics of a marine subtidal area extremely important for its natural resources. The Texas shelf can be described as a very dynamic system driven by a complex aggregation of meteorologic and oceanographic events, including diverse wind and current structures. Superimposed upon these phenomena are influences to the system both from the local rivers and estuaries as well as the distant Mississippi River and deep ocean waters of the Gulf basin.

The ecosystem represents a typical ocean environment in terms of nutrient concentrations and associated annual dynamics. The Texas shelf is relatively pristine concerning the pollutants monitored during this study, such as hydrocarbon and trace metal concentrations, with the majority of hydrocarbon observations during the study period being related to natural phenomena.

The shelf supports relatively high phytoplankton biomasses with extremely high annual production, especially in the inner-shelf waters. Many of the phytoplankton characteristics are strongly related to salinity and incident solar radiation as well as possible nutrient regeneration. Most of the marine biota observed in this study show strong geographical trends, usually related to water depth or distance from shore, as illustrated in Figure 52. The plankton are most abundant along the inner shelf, and numbers are predictably large in the spring, correlated to riverine input and nutrient maxima.

Ecosystem Characteristics

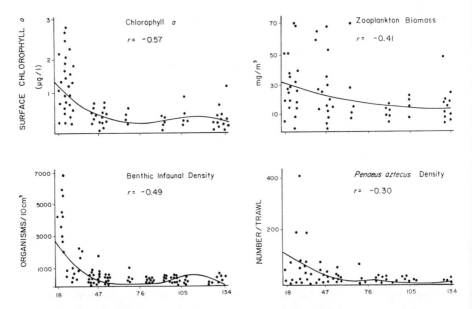

Figure 52. Relationship of chlorophyll *a*, zooplankton biomass, benthic infaunal density, and *Penaeus aztecus* (brown shrimp) density to water depth (m) for the south Texas outer continental shelf. Correlation coefficients (r) are indicated for each plot.

The inner-shelf maxima for phytoplankton were also reflected in zooplankton. Peaks in zooplankton biomass were observed at shallow sites with decreases occurring in an offshore direction. Both infaunal and epifaunal (as represented by *Penaeus aztecus* densities) organisms were more numerous along the inner shelf where general productivity was greater in response to increased food supplies. An additional factor potentially controlling the infaunal and epifaunal organisms appeared to be the coarser-grained sediments at the shallow sites. These sediments may provide a more suitable habitat than the finer silts and clays of the outer-shelf environment.

The larger and more mobile fauna, such as the demersal fish, showed less spatial patterning in their distributions on the shelf. The ichthyoplankton, however, which were strongly correlated to zooplankton, were far more abundant on the Texas inner shelf. It appears reasonable to conclude that the shallow areas of the south Texas shelf are biologically a more critical part of this marine ecosystem than are deep areas.

The changes in fauna observed suggest that the inner-shelf region is a much more dynamic area than the waters near the shelf break. In addition, much of the riverine input to the south Texas shelf enters through well-developed estuaries. These estuaries undoubtedly have an important impact on the shelf that may be manifested in some of the gradients illustrated in Figure 52.

The presence of isothermal conditions from the surface waters to the sea floor during much of the year allows for considerable interaction between two dynamic communities in the inner-shelf region of the Texas coast: (1) a benthic community consisting of those organisms living in or on the sediment or near the sediment-water interface and (2) a pelagic community consisting of those organisms drifting, floating, or swimming in the overlying waters. Because of their interactions, the boundaries of these two communities are not clear. Many nekton organisms, for example, deposit eggs that become part of the benthic community while the larvae and adults are members of the pelagic community. Conversely, numerous benthic species produce eggs that float in the water column, hatch into planktonic larvae, and become dispersed by currents before settling permanently to the bottom. In addition to these interactions, demersal fishes swim into the pelagic zone to feed on plankton, while the benthos depends upon the continual "rain" of materials (e.g., algae and fecal pellets) from the overlying waters for nourishment.

Nutrient Regeneration

Evidence from the three-year study indicates that the Texas shelf waters, especially inner-shelf waters, are extremely productive in plant biomass. Further evidence for this high production comes from the fact that several important commercial fisheries are supported in these same waters. Rowe et al. (1975) recently contended that nutrient regeneration in sediments is the major factor responsible for the relatively high rate of primary carbon fixation in continental shelf waters. They indicated that the lack of contributions of nutrients such as ammonia from bottom sediment would cause the loss of an important "feedback" to the system, which would leave the pelagic primary producers dependent solely on water column sources of nutrients. They speculated that if this were the case, shelf primary production would be reduced to rates observed beyond the shelf break.

The conclusions of Rowe and co-workers were based primarily on two observations. First, they observed gradients of decreasing ammonium (NH_4^+) concentrations between the benthos and overlying water column. Second, they estimated potential releases of ammonia by the sediments from measurements of respiration, assuming the oxidation of organic matter, including infaunal metabolism, would result in ammonia release.

More recently Carpenter and McCarthy (1978) attempted to shed doubt on the hypothesis of Rowe et al. by contending that on the continental shelf the sediments play a minor role in cycling nitrogenous nutrients for primary producers. Part of the basis for their contention was that primary producers in the water column of a shelf habitat do not have enough of their energy diverted to the benthos to cause rates of ammonia regeneration by the sediments as reported by Rowe et al., because much of the primary producers' energy is utilized by zooplankton and nekton on the shelf. Carpenter and McCarthy believed that for regeneration rates as described by Rowe et al. to be occurring, the sediments would have to receive supplemental supplies of allochthonous organic materials. As will be illustrated later in this chapter, data collected on the Texas shelf indicate that the majority of phytoplankton biomass produced on the STOCS inner-shelf waters does not go as energy to pelagic components of the system but rather is diverted directly to the benthos.

Data collected during several special cruises in the STOCS study, which were intended to trace the nepheloid layer dynamics on the Texas shelf, lend support to the original hypothesis put forth by Rowe and his colleagues. Figure 53 illustrates diel ammonium (NH_4^+) profiles through the water column at a station approximately 33 m deep on the Texas shelf during four periods in 1978. Almost every profile illustrates a gradient in ammonia through the water column with increases in concentration near the mud-water interface. These data were collected during conditions of water column stratification and nonstratification and showed similar trends during both. In addition, results presented in Chapter 3 (Figure 20) indicate that the peaks in bottom water nitrogen were often associated with chlorophyll peaks. In conjunction with these peaks, photosynthesis was shown to be occurring, suggesting that the nutrient concentrations are being utilized by primary producers.

Given the observations by Rowe et al. (1975) plus the experiences cited above for Texas shelf waters, there appears to be sufficient evidence to support the idea of sediment nutrient regeneration being at least partially responsible for higher rates of primary production in shelf waters. Under circumstances like these, the bottom may serve as a nutrient reservoir and may dampen the effects of surface productivity cycles. Furthermore, the occurrence of this phenomenon on

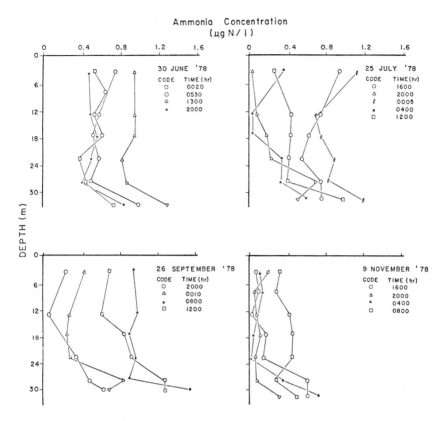

Figure 53. Temporal (by hour) profiles for ammonia nitrogen concentrations with depth during the sampling periods indicated at a sampling location near Station 4, Transect II.

the south Texas shelf emphasizes the importance of several other ecosystem mechanisms that should be briefly mentioned.

As stated above, the sediments of the STOCS ecosystem may act as a reservoir for nitrogenous compounds such as ammonia that are potentially usable for primary producers. Besides the apparent flux of ammonia out of the sediments, a flow generated by gradients between the sediments and overlying water, there is another possible mechanism for nutrient release on the Texas shelf that is directly related to the consistent occurrence of the nepheloid layer. As suggested in Chapter 3, the nepheloid layer is the result of sediment resuspension. The silty-mud nature of Texas shelf sediments help to perpetuate the presence of a nepheloid layer in these bottom waters. The benthic fauna, as well as macroepifaunal species that may disturb and otherwise bioturbate the bottom sediments, potentially serve as other influencing factors in the maintenance of this nepheloid layer with its associated nutrients, plant biomass, and detritus. The biological dynamics of the fauna in and on the sediments potentially provide aeration to the interstitial water and sediments as well as different degrees of substrate coherence and stability, dependent upon the type of biological function that occurs.

The recycling of nutrients as well as sediment detritus and their release to the water column depend largely on the ease with which the muddy sea floor can be resuspended. Bioturbation and current turbulence control this process (Rhoads, Tenore, and Browne 1974). Knowledge of the overall extent of biogenic activity is a key to partially predicting the consistency of nepheloid layer occurrence over various parts of the shelf. Furthermore, it is likely that the bioturbation activities of the benthic fauna play a role not only in the dynamics of the nepheloid layer but also in the general mechanisms responsible for nutrient regeneration across the mud-water interface. According to the evidence cited above and by Rowe et al. (1975), these nutrient regeneration dynamics of the Texas shelf could serve as a major force driving the ecosystem and responsible for some of the extremely productive fisheries supported by this shelf.

Trophic Coupling

For many years immense amounts of information have been accumulating on primary production, zooplankton abundance, and the distribution of benthic organisms in important fishing areas. Despite these data bases it is very difficult to describe quantitatively the links between primary production and fish yields. A few plausible attempts to quantify these links have been provided by Steele (1974) for the North Sea ecosystem and by Mills and Fournier (1979) for the Scotian shelf system. Even without complete data bases, the comparison of regions like the North Sea, the Scotian shelf, and, for example, the northwestern Gulf of Mexico shelf should offer insight into the general structure of marine ecosystems and pinpoint deficiencies in our understanding of them. Of most concern here is the need to take a hard look at the hypothesis that, despite geographical differences, most coastal ecosystems with productive fisheries have similarly constructed food webs (Dickie 1972; Mills 1975).

Ecosystem Characteristics [143]

The importance of understanding the functioning of an ecosystem, especially in relation to an important fishery, cannot be overemphasized. For example, to demonstrate the effect of a perturbation such as an oil spill on an ecosystem and to relate that directly to an impact on man, the effect on a natural resource such as a fishery must be cited. This effect on fishery productivity as reflected by commercial catch statistics is risky at best. There are several shortcomings in fishery catch statistics. They do not represent precise reporting because they fail to take into account the changing effort and technological advances of fisheries. They also are generally not available for the localized areas in which the perturbation may be intense.

Based on these shortcomings and the desire to better understand the components of an important resource to the Texas shelf, we decided to use both the STOCS data base as well as bibliographical information to derive a conceptual model of the trophic relations involved in the shrimp fishery on the shelf. Through correlation research, we sought relationships between different components of the STOCS data base that intuitively made biological sense. The result of this search was the development of a conceptual model, based on correlation coefficients, that suggested relationships among the water column, benthos, and shrimp that could serve as the basis upon which to build a food web hypothesis.

These relationships are depicted in Figure 54. Only significant correlation coefficients ($P < 0.05$) are illustrated. The model emphasizes several patterns. There is a relationship between the water column fauna, in this case zooplankton, and the sediment detrital pool, illustrated by the correlations between zooplankton nickel body burdens and sediment nickel concentrations as well as several zooplankton hydrocarbon body burden variables and hydrocarbons observed in the sediment. The hypothesis that could be derived from these results is that zooplankton fecal pellets serve as a major input to marine sediment detrital pools. This has been verified by numerous studies as referred to in Steele (1974).

In addition to zooplankton input to the benthos, the data indicate that primary producer biomass, as represented by bottom water chlorophyll concentrations, is related to densities of benthic infauna and bacteria, potentially through the detrital pool. Furthermore, the relationships depicted by the model suggest the suspected relation between sediment hydrocarbon concentrations and bacteria with the former serving as a potential food source.

The interrelationships that are suggested in Figure 54 for the various size categories of fauna living within the sediments (bacteria,

[144] Ecosystem Characteristics

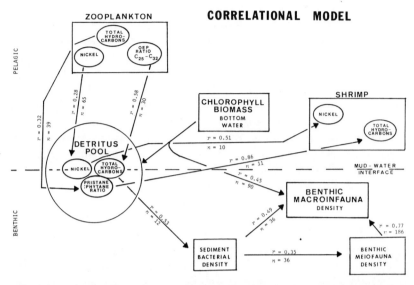

Figure 54. Schematic representation of significant ($P < 0.01$) correlations found among the STOCS variables indicated. The correlation coefficients (r) and number of cases (n) are also shown.

meiofauna, and macrofauna) are relatively strong and provide further insight into the functioning of the Texas shelf benthos. The constant ratio of benthic animals to bacteria, and not organic carbon, indicates that benthic animals are related more to bacteria than to organic carbon. This relationship suggests that benthic animals utilize bacteria and not organic carbon as a food source. Bacteria are considered a major food item for a wide variety of meiofauna (Coull 1973) and macrofauna (Zobell and Feltham 1938; Newell 1965; Chua and Brinkhurst 1973).

There are some indications that the meiofauna may contribute more to the matter and energy cycles of the sea than was envisaged by earlier investigators (Perkins 1958; McIntyre 1969). Recent caging experiments (Bell and Coull 1978; Rubright 1978; Buzas 1978) have shown that meiofaunal populations, particularly the Nematoda, the Foraminiferida, and the Polychaeta, are substantially reduced by predation and, therefore, probably represent an important food source. Buzas (1978) performed predator exclusion experiments in which foraminiferal biomass inside meiofaunal cages were 3 to 12 g/m² higher than outside the cages. This suggested foraminiferal densities were significantly reduced by predation. A study by Bell and Coull (1978) indicated that *Palaemonetes* (grass shrimp) predation and disturbance significantly lowered total meiofaunal densities and that the shrimp randomly fed on the available nematodes in proportion to the nematodes' abundance.

Macrofauna acting as surface deposit feeders can shift to subsurface deposit feeding when high quality sedimentated food following a spring phytoplankton bloom diminishes. Others are exclusively subsurface deposit feeders. Much meiofaunal predation may be incidental to nonselective subsurface deposit feeding. In the sediment, the organic food is more refractive, and the energy source available to macrofauna and meiofauna is via bacteria and organic fractions that can be digested (Gerlach 1971).

Gerlach (1978) further states that although the estimates of meiofaunal contributions to the organic content of sediment utilized by subsurface deposit feeders are considered low, the bacterial biomass is not very much higher than meiofaunal biomass. He states that meiofauna must be considered an important food source if the concept of nonselective feeding is valid.

Similar complex feeding modes are found within the meiofauna. With increased study of the life histories of meiofauna, as with that of the macrofauna, the classical consensus that meiofauna are detrital feeders or indiscriminant feeders on benthic diatoms and bacteria (Wieser 1960) has been replaced with the idea that they show as varied a feeding mode and diet as exists in the sediments (Coull 1973). Some meiofauna are active predators. Many feed on bacteria and protozoa in competition with macrofauna, and others assimilate dissolved organic matter directly.

The primary function of meiobenthos in trophic relationships has traditionally been the assistance of recycling nutrients at a low trophic level (McIntyre 1969). However, added importance is now being attributed to meiofauna in the enhancement of microbiota environment (growth) on detritus. Growth of an associated microbiota is enhanced when the detrital feeder mechanically breaks down the particle and increases the surface-to-volume size, thus furthering the microbial growth and decomposition (Coull 1973). There is little doubt that particulate organic detritus is consumed in great quantities by meiofauna. Gerlach (1978) stated that in situ sediment bacterial production is far below potential rates and somewhat stationary. Furthermore, he found that faunal activities may be beneficial for bacterial growth in marine sediments and may stimulate the rate of detritus decomposition. This statement applies to microfauna as well as meiofauna, which would place both categories back into the same complex food web. Meiofauna may be more efficient, however, in utilizing organic substrates than macrofauna (Gerlach 1978).

Gerlach (1978) reported more specific relationships between meiofauna and sediment bacteria. In microcosm experiments, the breakdown of ^{14}C-labeled *Zostera* detritus was greater when meiofauna was present than when there were only polychaetes. Nematodes may share with protozoa a major role in benthic nutrient regeneration and prevent bacteria from reaching self-limiting numbers. More specific interactions between meiofauna and bacteria were also reported by Gerlach (1978), including one in which the cuticles of certain nematodes were covered with a sheet of densely packed bacteria that they fed on and "gardened" by providing favorable environmental conditions, for example, by migrating up and down between aerobic and sulfide layers of the sediments. Other simple associations such as "mucus traps" on nematodes to which organic particles adhere resulting in a subsequent bacterial growth may be more widely distributed in marine sediments than is now known.

Finally, Figure 54 depicts a relationship that potentially ties the density of shrimp on the Texas shelf to the functioning of other major components of the ecosystem. There are strong correlations shown for shrimp body burdens of nickel and total hydrocarbons with sediment nickel concentrations and a hydrocarbon variable, suggesting that shrimp may derive their nutrition from the benthos. More important, however, are the relationships that are portrayed for nickel concentrations throughout the model depicted in Figure 54 (zooplankton ▶ sediment ▶ shrimp). These correlations make it possible to propose a trophic coupling hypothesis for shrimp that includes both pelagic and benthic components. It is quite clear that in inner-shelf waters, where mixing occurs resulting in a relatively homogeneous water column, the discrimination between pelagic and benthic components is very obscure, and the potential for trophic coupling between the two becomes very important.

The subtidal region near shore of the Texas coast with its many interacting communities is the site of several major fisheries including penaeid shrimp. As a result of the south Texas shelf correlational model detailed in Figure 54, we feel it is imperative to examine some of the trophic interactions of this region and relate them to a fishery of immense economic importance in order to delineate the deficiencies in our understanding.

Outside the bays and estuaries, the shrimp fishery extends to approximately 80 m depth on the shelf, with maximum yields obtained well inside this range. Annual shrimp landing reports, National Oceanic and Atmospheric Administration National Marine Fisheries Service Gulf Coast Shrimp Data, Annual Summaries, indicate that for the reporting area (Statistical Area #20) similar to STOCS stations monitored 1975–1977 (Figure 55), an annual average of 5.7×10^6 kg of shrimp were landed for the years 1975–1976. This represented a mean value of $18 million for that period to commercial fishery.

For purposes of developing a conceptual model, a single station centered in the middle of the fishery reporting area described above, which was monitored on almost a monthly basis 1976–1977, will be emphasized. This station, Station 1/II (reference station, Figure 55) was located off Aransas Pass Inlet at a depth of approximately 22 m.

Primary production for Texas inner-shelf waters as characterized by the reference station was somewhat bimodal on an annual basis with peaks in the spring and fall (Chapter 3, Figure 17). Annual estimates of production based upon chlorophyll a measures converted to carbon equivalents, according to the methods of Ryther and Yentsch (1957), indicated that these waters produced a mean of approximately 103 gC/m^2/yr (Figure 56).

Macrozooplankton biomass on the Texas shelf averaged approximately 3.566 g/m^2 wet weight over the sampling interval. If we assume a turnover ratio of seven (Steele 1974), annual production of the macrozooplankton was estimated to be 25 g/m^2/yr. Since the water column was usually fairly homogeneous and the zooplankton tows often did not reach the bottom (and if sampling was biased from net clogging), it is likely that the estimate for production should be doubled to 50 g/m^2/yr for purposes of this model. Assuming approximately a 6% conversion between wet weight and carbon content of metazoans, we can estimate the carbon equivalent of zooplankton production to be 3 gC/m^2/yr (G. Rowe 1979, personal communication).

Information on the neuston component of the planktonic community indicated that an additional 0.21 gC/m^2/yr could be assumed for the macroplankton production from these surface animals. Standing crop of microplankton was calculated to be 465 mg/m^2 wet weight. Annual production was estimated as 10 times the standing crop because of larger expected turnover ratios for the microplankton. With the conversion to carbon content mentioned above, this resulted in approximately 0.9 gC/m^2/yr; therefore, the total production estimate for the zooplankton component of the food web on the Texas inner shelf is approximately 4.1 gC/m^2/yr.

If we assume a very conservative minimum transfer efficiency of 20% between primary producers and the zooplankton, then 20.5 gC/

[148] Ecosystem Characteristics

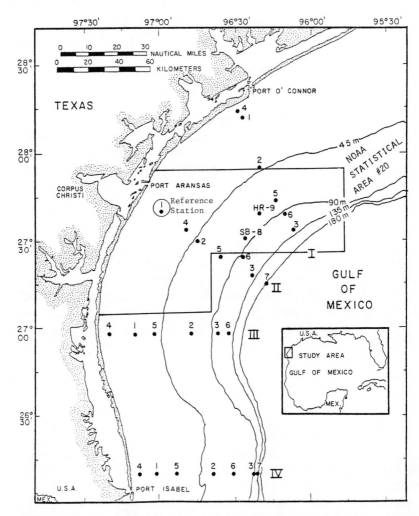

Figure 55. Location of the reference station focused upon in the NOAA statistical reporting area (L-shaped area) for the development of a trophic model for shrimp.

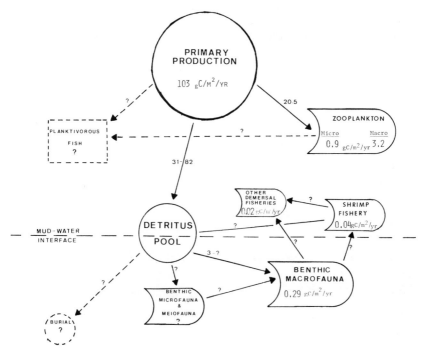

Figure 56. Conceptual trophic model for the brown shrimp fishery. Material transfers indicated are also in $gC/m^2/yr$.

m^2/yr would be required to support the zooplankton. This transfer of carbon results in approximately 82 $gC/m^2/yr$ of primary production remaining. Mills and Fournier (1979) indicated that, in contrast with that of the North Sea ecosystem (Steele 1974), the majority of primary production for the coastal ecosystem on the Scotian shelf was diverted to the demersal fisheries. This may very well be the case for the Gulf coastal ecosystem also. The bottom waters appear to support greater amounts of primary producers than the surface or mid depths during most of the time (Chapter 3, Figure 13).

The amount of pelagic fisheries biomass that is directly supported by primary producers on the Texas inner shelf is unknown. From the amount of zooplankton production observed, however, one would have to assume that the pelagic fisheries are small. Therefore, the Texas inner-shelf ecosystem is probably characterized as a system in which the majority of primary production is input directly to the bottom waters and benthos.

[150] Ecosystem Characteristics

Information from Steele (1974) indicates that 30% of the primary production is transported to the benthos in the North Sea ecosystem. If we consider the above facts and assume there are no other major links to pelagic fisheries other than through zooplankton, it would appear that almost 80% of this production reaches the benthos in the Texas coastal waters. This is probably an overestimation, but the real number is certainly greater than the 30% estimated for the North Sea.

Further illustrating the input to the bottom, data from several cruises to examine nepheloid layer dynamics (which were detailed in Chapter 3) substantiate the presence of peak chlorophyll layers in these bottom waters. The carbon production at this depth plus the direct input to the benthos of detritus, both from the nepheloid layer and the upper portions of the water column, presumably can provide a fairly plentiful nutritional source for demersal-oriented trophic links.

Estimates of benthic infaunal biomass in this region of the Texas inner shelf range between 0.5 g/m^2 (STOCS study) and 2.5 g/m^2 (Rowe, Polloni, and Horner 1974) (Table 24). Assuming a turnover ratio of approximately 4.5 (Nichols 1978), an average of 0.29 gC/m^2/yr are produced by the infaunal benthos (Figure 56).

Shrimp fisheries yields (NOAA/NMFS Gulf Coast Shrimp Data, Annual Summaries) were used to estimate the production of shrimp on an annual basis for the inner-shelf waters. Utilizing the suggested conversions to obtain the heads-on weight and assuming a turnover ratio of approximately 0.8, the commercial fishery catch represented approximately 0.03 gC/m^2/yr of shrimp production (C. W. Caillouet 1979, personal communication). According to the hypothesized survival curve presented in Figure 57 this estimate of shrimp production was for approximately 78% of the shelf population in NOAA's Statistical Area #20 (Figure 55). Therefore, adding the other 22% of the population, annual production was estimated to be 0.04 gC/m^2/yr (Figure 56).

Data from the STOCS study indicated that an additional 0.02 gC/m^2/yr of other demersal species were produced on the inner shelf. The combination of this data with the shrimp production estimates illustrated that approximately 0.06 gC/m^2/yr was produced by the fauna living in these bottom waters of the Texas shelf. Comparing this trophic level to the infaunal production and assuming a standard 10% transfer efficiency, we see that benthic biomass is insufficient as a food source to support solely the demersal component of the inner-shelf food web. These figures do not include meiofaunal production, but even if this component were known, there probably still would not be enough biomass to support directly the demersal fisheries.

Furthermore, as illustrated in Table 24, there appears to be much less benthic infaunal production in the northwestern Gulf of Mexico than in other continental shelf regions such as the northwestern Atlantic. This is surprising considering the extensive fishery supported on the Gulf continental shelf.

The alternative to an infaunal-demersal fishery trophic link is a detrital-based trophic web for many of the commercially important species, including the shrimp. The data on primary production plus the peak concentrations of chlorophyll in the bottom waters along with a relatively small amount of pelagic secondary production would tend to support this conclusion.

If the Texas inner shelf trophically revolves around a detrital food web, one of several questions to ask concerns where the benthos fit into this trophic scheme, especially since they do not appear to have the biomass to support alone the observed production at higher trophic levels. A possible hypothesis for the role of the benthos takes into account the dynamics of the nepheloid layer. Rhoads, Tenore, and Browne (1974) pointed out that the concentration of suspended solids in many estuaries and coastal waters is higher in the bottom waters than at the surface, especially where the water column passes over muds that have undergone intensive bioturbation.

As stated previously, the recycling of materials, such as detritus, from the sediment to the bottom waters depends largely on the ease with which the muddy sea floor can be resuspended. Bioturbation by infauna and current scour control this process (Rhoads, Tenore, and Browne 1974). Primary productivity in turn provides plankters to the bottom waters through surface sedimentation. Both living and dead plankters plus associated microorganisms produce detrital food for demersal consumers including shrimp populations. Thus, benthic infauna do not necessarily provide all of the direct food sources for an important fishery such as shrimp but rather supplement the demersal consumer's diet and indirectly provide alternative nutritional sources through their bioturbation activities and the maintenance of a very productive zone in the near-shelf bottom waters.

In turn, the extremely high densities of shrimp on the Gulf of Mexico shelf, as indicated by the successful fishery, probably have a direct effect on the smaller benthic infaunal biomasses observed for these waters as contrasted with those of the Atlantic coastal waters (Table 24). The predation pressure of the shrimp plus their physical feeding activities may serve as influential factors in maintaining infaunal organisms at relatively smaller sizes with possibly higher turnover ratios than even assumed here.

From the preceding exercise it is obvious that the coastal waters of

[152] Ecosystem Characteristics

Table 24. Comparison of Abundance of Biomass of Macrobenthos from the Northwestern Atlantic Ocean and Northwestern Gulf of Mexico

	Atlantic Ocean[a]	
Depth (m)	Density (No./m²)	Wet Weight (g/m²)
30	26,060	7.69
40	7,390	2.44
Avg	16,725	5.07

[a] Measures from Rowe et al. (1974).
[b] Measures from the South Texas Outer Continental Shelf Study, 1975–1977.

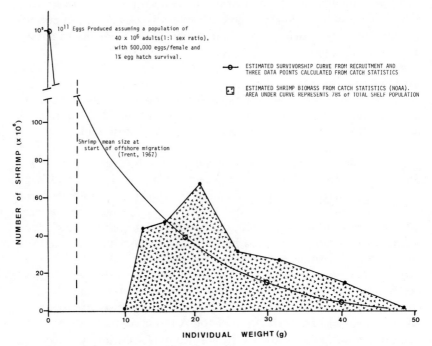

Figure 57. Plot of the reported shrimp fishery yield (shaded area) according to size range along with an estimated survivorship curve (solid line) for the south Texas shelf brown shrimp population.

	Gulf of Mexico			
Depth (m)	Density^a (No./m^2)	Wet Weight^a (g/m^2)	Density^b (No./m^2)	Wet Weight^c (g/m^2)
12			1,536	0.63
16	1,373	0.74		
30	14,623	4.09	675	0.28
	7,998	2.42	1,106	0.46

^{c}Wet weight calculated from densities of organisms using the density–to–wet weight ratio of the respective values from Rowe et al. (1974).

the Gulf of Mexico are extremely productive and that this production is influenced by many factors. Much of this production appears to be diverted directly to the benthos, and the objects of major regional fisheries, such as shrimp, appear to receive much of their nutrition from a detrital food web. Determining the mechanisms of this food web and the exact role of such components as the benthos is an extremely important task for future research. This ecosystem with its food webs leading to major commercial fisheries appears certainly different in structure from, for example, the system described by Steele (1974) for the North Sea. This difference points to the need for completing detailed regional studies before making generalizations and constructing models to predict the effects of such factors as environmental disturbance on important fisheries.

Environmental Disturbance

In the state of Texas, one third of the population resides in the coastal zone. A total population of 3.5 million was recorded for the state in 1975, and the population was projected to increase to 5 million by 1980. Such dynamic growth in Texas and throughout the United States implies parallel expansion in the coastal zone and increased demands for manufacturing; petroleum and natural gas exploration, production, and refining; marine transportation; commercial use of natural resources; and recreation and tourism.

The pressures imposed by a rapidly increasing population demand intelligent control and management of coastal regions. The coastal waters near shore make up less than 1% of the world's oceans, yet it is in this fringe that the most productive ecosystems in the world are found. The different water masses influencing the Texas shelf, in particular freshwater discharge, suggest that as salinity decreases from riverine input the particulate matter increases along with possible associated nutrients and primary productivity. The effects of this scheme are felt throughout the south Texas shelf. Thus, the outer continental shelf system is not just a product of the dynamics of Gulf waters but a reflection of many processes in coastal waters near shore as well. Our continued multiplicity of demands upon the complex coastal environments make it imperative that their functioning and ecological values be understood. This knowledge is essential for decision makers to properly manage coastal environments while maintaining the best possible conditions for the continued productive uses of natural resources.

Healthy, naturally functioning ecosystems are one of the principal resources in the northwestern Gulf of Mexico that are susceptible to environmental disturbance, such as oil spills. The benefits we reap from these systems include multimillion dollar commercial fisheries, both of shellfish and finfish, and particularly of penaeid shrimp. The health of the general public can be threatened by contaminated foodstuff passing through the fishery markets. A sizable sports fishery also harvests these resources. An important segment of our coastal economy is based on recreation and tourism that in turn are dependent upon the natural resources. Also, segments of the shelf ecosystem provide habitats for endangered and threatened species, such as sea turtles. Billions of dollars per year are contributed to our economy from the northwestern Gulf of Mexico as a result of all these fish and wildlife populations.

In addition, the Gulf coastal zone must be considered one of the most critical nonliving resources. Population centers naturally develop along the coast and serve as focal points for national and international commerce as well as recreation. Wastes of various kinds normally enter Gulf waters by direct discharge from coastal municipalities and industries or through tributary streams that serve as transmission media for waters from larger areas. Included in this waste are raw and partially treated municipal sewage, industrial wastes including petroleum products, and sediment loads from soil erosion. In an era of increased energy demands, the coastal zone is an area that may be affected by oil and gas exploration and production, oil spills, increased marine transportation, and manufacturing and industrialization. Potential exists for increased sources and amounts of pollutants to enter the system. Impacts on these areas from an environmental disturbance may include loss of income from decreased fisheries, impeded tourism or recreation or both, created health dangers, and damaged natural resources.

The objectives of the three-year study of the south Texas outer continental shelf were to give descriptions of the physical, chemical, and biological components of the system and their interactions against which subsequent changes or impacts could be measured, particularly concerning the effects of outer continental shelf oil and gas development activities. The synthesis and integration of this data were designed to develop an encompassing description of the study area, identifying the temporal and spatial trends that best represented the ecosystem along with mathematical descriptions for unique relationships that would serve as "fingerprints" for future comparisons.

The study results indicate that a major step has been taken toward reaching this objective. Our analyses have shown the south Texas shelf to be an extremely productive and complex system. Meaningful relationships have been found between the physical, chemical, and biological components studied, and naturally inherent variability has been quantified. Potential, implicative, or biologically intuitive relationships have been hypothesized that describe some of the forceful dynamics of this system, but these exercises only serve to point out the need for detailed studies before making overall generalizations and constructing predictive models.

Evidence from the study indicates that the south Texas shelf bottom environment is relatively pristine in regard to those chemicals monitored during the study period. There is minimal petroleum pollution of sediments (lack of aromatic hydrocarbons), and hydrocarbons present appear to be free of any significant trace metal contamination. Likewise, the water column is relatively pristine in regard to the hydrocarbons studied, and researchers attribute those observed during the study primarily to natural sources such as suspected natural gas seeps along the edge of the shelf. Because of the tremendous quantities of available dilution water from local and Mississippi riverine input, pollution wastes originating in the coastal area have had minimal effects upon either water quality or resources in the Gulf of Mexico. There is, however, increasing pressure to further develop many of the energy resources in the Gulf, development that can severely threaten the pristine nature of the waters, especially in terms of hydrocarbons.

The ecological effects of oil pollution on the Texas shelf environment are important considerations in energy policy decisions affecting this area, primarily because of economic factors such as the extensive commercial fisheries. At present, assessment of the environmental impact of energy resource development must be made somewhat in ignorance and uncertainty because of conflicting opinions and lack of knowledge. The only remedy for our uncomfortable lack of knowledge is programs such as the one detailed in this presentation. We hope that the ecosystem description presented here will begin to close the wide gaps in existing information. As we have seen, the Texas shelf is a complex collection of many components, all interrelated and interdependent for the well-being, health, and productivity of the whole system. Although this study is a beginning, we are still a long way from the level of commitment required to completely understand the Gulf ecosystem and the effect on the environment from the unnatural impacts of oil and gas exploration and production.

APPENDIX A
OVERALL BASE LINE RESULTS

Distributional Characteristics of Selected Important Variables[1]

The purpose of this appendix is to present base line information for the south Texas outer continental shelf (STOCS) for 1975–1977. Presented are the distributional characteristics (i.e., mean, standard deviation, skewness, kurtosis, and confidence intervals) for variables selected from each of the study areas of the STOCS study integration effort. Also presented is information concerning significant temporal and spatial variation for these variables.

Tables A-1 through A-4 explain the actual numerical results that are presented in Table A-5 at the end of this appendix. Preceding Table A-5 are several explanatory sections: (1) a section describing the format and use of Table A-5; (2) an overview of the sampling scheme of the STOCS study; (3) a methodology section describing the selection of variables and the statistical methods used; (4) a section discussing the comparison of the present base line results with future monitoring results; and (5) a section discussing the number of samples needed in future monitoring efforts. No attempt has been made in this appendix to discuss the scientific meaning or implications of the various numerical results. Such discussion has already been presented in the main text. This appendix is intended to assist future workers in efficiently locating desired base line results and in quickly comparing these results with values from future monitoring efforts.

Format and Use of Table A-5

Table A-5 has been organized to present several different types of results, and the format is somewhat complicated. A set of hypothetical

Appendix A

Table A-1. Hypothetical Results Illustrating Table A-5 Format

Variable	Mean	SD^a	Skew.[b]
High molecular-weight hydrocarbons			
EPIFAUNA (1976–1977)			
Rhomboplites aurorubens			
Liver			
Total Hydrocarbons (µg/g)	13.60	9.78	0.84
($n-C_{14}$ to $n-C_{32}$)			
Station 1	13.04	8.79	0.73
Station 2	15.96	9.30	0.74
Station 3	11.26	10.64	0.92
1976	10.68	9.63	0.80
1977	14.91	9.43	0.85
Pristane:phytane	81.41	59.10	0.79
Pristane:$n-C_{17}$	10.67	9.92	1.65
Winter	7.03	—	—
March	6.31	2.98	−0.49
April	14.97	7.63	2.91
Spring	20.60	10.42	3.41
July	5.78	10.80	1.56
August	5.29	6.99	1.72
Fall	8.46	9.67	1.82
November	2.18	9.56	1.84
December	1.85	—	—
Phytane:$n-C_{18}$	0.68	0.52	0.46

[a] SD, standard deviation.
[b] Skew., skewness.
[c] Kurt., kurtosis.
[d] Empir. CI, empirical confidence interval.
[e] N, number of cases.

results is presented in Table A-1 that will serve to illustrate the general format of Table A-5. The column titles in the table refer to the distribution statistics presented for each variable. The actual methods used to calculate the different distribution statistics are spelled out in the methodology section of this appendix.

In Table A-1, "High molecular-weight hydrocarbons" and "Epifauna" indicate the study area for the variables that follow. In parentheses after the study area title are the years for which data are being considered, 1976–1977 in this example. Note that the variables that follow are relevant to a specific species (*Rhomboplites aurorubens*) and a

Kurt.[c]	95% Empir. CI[d]	N[e]
0.08	0.57– 35.80	50
0.01	0.57– 32.91	20
−0.05	2.98– 35.80	20
0.18	6.29– 28.76	10
0.07	5.43– 29.78	15
0.08	0.57– 28.76	35
−1.34	17.27–171.60	40
2.02	1.85– 36.00	42
—	7.03– 7.03	1
—	3.85– 8.22	3
1.42	5.94– 36.00	5
2.12	6.92– 28.72	5
1.92	1.85– 16.92	14
1.62	4.93– 6.77	4
2.60	6.91– 10.32	5
2.11	7.12– 11.92	4
—	1.85– 1.85	1
−1.60	0.17– 1.50	46

specific tissue (liver). The first subtopic for this species-tissue combination is "Total hydrocarbons." This variable is the sum of the normal alkanes involving from 14 to 32 carbon atoms ($n-C_{14}$ to $n-C_{32}$), and the units for this variable are micrograms per gram (µg/g). There was a total of 50 cases considered for the "Total hydrocarbon" variable; the overall mean was 13.60, and the overall standard deviation, 9.78, and so forth. Note that "Total hydrocarbons" is then broken down by station and then by year. For example, the 20 data cases for Station 1 had a mean of 13.04 and a standard deviation of 8.79, while the 15 data cases for 1976 had a mean of 10.68 and a standard deviation of 9.63.

[160] Appendix A

The fact that breakdowns by station and by year are presented for "Total hydrocarbons" signifies that valid significant differences (beyond the 0.05 level of significance) were found among the three stations and also between the two years. In general, a series of analyses of variance[2] were performed to test each variable for significant temporal and spatial differences—differences due to transect, station, collection period (month or season), and year. If an effect (e.g., transect) was consistently significant for a variable (i.e., significant in all analyses involving that effect) and if that effect was general (i.e., there were no significant interactions involving that effect), then valid significance was accorded that effect, and a breakdown of the effect was included in the table. This system for selecting valid significant results is quite conservative and has limited the significant temporal and spatial effects reported in Table A-5 to those that are general and unambiguous. Further details are presented in the methodology section.

Again consider the hypothetical results in Table A-1. Note that collection period and transect are not broken down for "Total hydrocarbons," indicating that valid significant differences were not found for these two effects. The second variable in Table A-1 is the pristane-to-phytane ratio ("Pristane:phytane"). The lack of breakdowns for this variable indicates that no valid significant temporal or spatial differences were found. Note that no units of measurement are presented for "Pristane:phytane" since it is a unitless ratio.

The third variable in Table A-1 is the pristane–to–n-heptadecane ratio ("Pristane:$n-C_{17}$"). The collection period has been broken down for "Pristane:$n-C_{17}$," indicating valid significant differences for that effect. Note that standard deviation, skewness, and kurtosis values are not present for winter or December. These statistics were not calculable because there was only one data case for each of these two periods. Also note that no kurtosis value is presented for March. There were only three data cases in March, and a minimum of four data cases is required for the calculation of kurtosis. In general, blank values indicate that the statistic could not be calculated.

In the "95% Empir. CI" column in Table A-1, there are two values separated by a dash. The value preceding the dash is the lower limit of the confidence interval, and the value following the dash is the upper limit of the confidence interval. Note that the lower limit of the 95% empirical confidence interval corresponds to the 2.5 percentile of the observed distribution, and the upper limit of the confidence interval corresponds to the 97.5 percentile. Such an empirically defined 95% confidence interval can be quite different from the theoretically

Overall Base Line Results [161]

defined 95% confidence interval (mean ± 1.96 × standard deviation) so familiar in the literature. The theoretically defined confidence interval is based on the assumption of an underlying normal distribution. Such an assumption of normality is not valid for a large number of the variables presented, thereby obviating the general use of the theoretical confidence interval. Further discussion of both empirical and theoretical confidence intervals is presented in the methodology section of this appendix.

Overview of the Sampling Scheme

The variables presented in Table A-5 represent several different sampling schemes. For most variables, data were collected for all three years of study (1975–1977). There are exceptions, however, with data being collected in only one or two years for some variables. In some cases, the principal investigator for a study area had questions about the validity or reliability of a variable for a particular year. In such cases, those data for the year in question have not been considered in the construction of Table A-5.

Two different sampling schemes were employed for collection periods. Some variables were sampled three times a year (winter, spring, fall); this scheme was referred to as seasonal sampling. Other variables were sampled nine times a year (Winter, March, April, Spring, July, August, Fall, November, December); this scheme was referred to as monthly sampling. Spring collections occurred in May and June. Fall collections usually occurred in September and October. (One of four fall cruises in 1975 was made in August.) Winter collections usually occurred in January and February. (Two of six cruises for 1975 were made in December 1974, and one of seven winter cruises for 1976 was made in March 1976.)

Spatially (geographically) three different sampling schemes were employed for the total study area as shown in Figure A-1: (1) a 12-station scheme involving Transects I through IV, primarily for water column (pelagic) sampling; (2) a 25-station scheme involving Transects I through IV, primarily for benthic sampling; and (3) a 2-station scheme involving one station on the Southern Bank (SB) and one station on Hospital Rock (HR).[3] For the 12-station scheme, stations were classified into one of three groups on the basis of depth (Table A-2). Variables collected according to the 12-station scheme were analyzed for two spatial effects—station group (1–3) and transect (I–IV). For the 25-station scheme, stations were classified into one of six groups on the basis of depth (Table A-3). Variables collected accord-

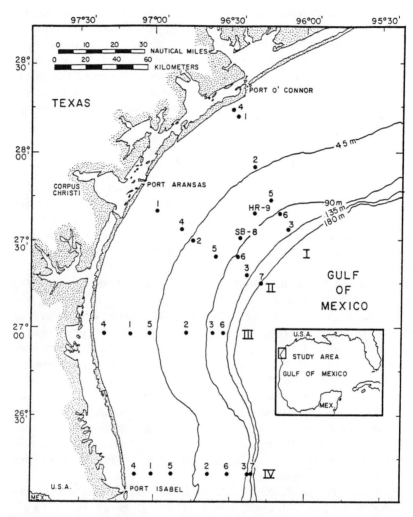

Figure A-1. Sampling sites for the STOCS study. The 12-station (pelagic) scheme involved Stations 1, 2, and 3 on Transects I and IV. The 25-station (benthic) scheme involved Stations 1–6 on Transects I–IV and Station 7 on Transect IV. The single station marked "HR-9" refers to Hospital Rock, and the station marked "SB-8" refers to Southern Bank.

Table A-2. Stations Grouped by Depth for the 12-Station Sampling Scheme

Station Group	Depth Range (m)	Transect	Station	Depth (m)
1	18–27	I	1	18
		II	1	22
		III	1	25
		IV	1	27
2	42–65	I	2	42
		IV	2	47
		II	2	49
		III	2	65
3	91–134	IV	3	91
		III	3	106
		II	3	131
		I	3	134

ing to the 25-station scheme were analyzed for two spatial effects—station group (1–6) and transect (I–IV). Variables collected according to the 2-station scheme were analyzed for a single spatial effect, SB compared with HR.

Methodology Used in Constructing Table A-5

SELECTION OF VARIABLES A variable was selected for inclusion in Table A-5 if (1) the principal investigator for the study area in question recommended it as an important base line characteristic or (2) that variable was found to have significant temporal or spatial variation. That a variable demonstrated temporal or spatial differences indicated that it was sensitive to environmental changes and thus a possible candidate for future monitoring.

DESCRIPTIVE STATISTICS For many variables, replicate samples[4] were not taken consistently and were therefore scattered over the different sampling sites and times. To allow a uniform approach to all variables, data from replicate samples were averaged to arrive at a single mean data case for each site-period-year combination. The number of data cases (N) in Table A-5 refers to the number of such mean values, and all descriptive statistics were calculated on the basis of these mean values. The mean ($\bar{X}$) is self-explanatory, indicating the normal

Table A-3. Stations Grouped by Depth for the 25-Station Sampling Scheme

Station Group	Depth Range (m)	Transect	Station	Depth (m)
1	10–18	I	4	10
		III	4	15
		IV	4	15
		I	1	18
2	22–27	II	1	22
		III	1	25
		IV	1	27
3	36–49	II	4	36
		IV	5	37
		III	5	40
		I	2	42
		IV	2	47
		II	2	49
4	65–82	IV	6	65
		III	2	65
		II	5	78
		I	5	82
5	91–106	IV	3	91
		II	6	98
		I	6	100
		III	3	106
6	125–134	III	6	125
		IV	7	130
		II	3	131
		I	3	134

arithmetic average. The standard deviation (SD) presented in Table A-5 is the unbiased estimate of a population value given by the following expression:

$$SD = \left[\frac{\sum_{i=1}^{N} (X_i - \bar{X})^2}{N-1} \right]^{1/2}.$$

In the above expression, N refers to the number of data cases and X_i refers to the value for the ith data case.

A basic characteristic of a distribution is skewness (Skew.). Skewness is a measure of the extent to which a distribution is symmetric about its mean. The measure of skewness presented in Table A-5 is calculated according to the following expression:

$$\text{Skew.} = \frac{\sum_{i=1}^{N} (X_i - \bar{X})^3}{N(SD)^3} .$$

If the skewness value is 0, then the distribution is symmetric. If the value is positive, then the tail to the right of the mean is drawn out more than the tail to the left. The converse is true for negative skewness values—the tail to the left of the mean is drawn out more than the tail to the right. An important use of a measure of skewness is determining if a distribution is normal in shape or not. A normally distributed population will have a skewness value equal to 0, and samples drawn from that population will have skewness values close to 0. Tables listing critical values for testing skewness, as illustrated in Snedecor and Cochran (1967, p. 552), can be used to test whether an obtained skewness value is significantly different from 0. If the absolute value of an obtained skewness exceeds the reported value for the appropriate sample size and desired significance level, then that skewness is significantly different from 0.

Another characteristic of a distribution is kurtosis (Kurt.). Kurtosis is a measure of the relative peaking or flattening of a distribution. The measure of kurtosis presented in Table A-5 is based on the following expression:

$$\text{Kurt.} = \frac{\sum_{i=1}^{N} (X_i - \bar{X})^4}{N(SD)^4} - 3.$$

A normal distribution will have a kurtosis of 0. If the kurtosis is positive then the distribution is more peaked (narrow) than would be true for a normal distribution, but a negative value means that it is flatter. A table of critical values for testing kurtosis, as illustrated in Snedecor and Cochran (1967, p. 552), can be used to test whether an obtained kurtosis value is significantly different from 0 (i.e., whether the distribution deviates from normality). To test kurtosis, simply enter the critical value table with the appropriate sample size and the desired level of significance. A negative kurtosis value is significantly different from 0 if it is less than the negative kurtosis value shown; a positive kurtosis is significantly different from 0 if it exceeds the positive kurtosis value shown.

[166] Appendix A

The confidence intervals presented in Table A-5 are 95% empirical confidence intervals (95% Empir. CI). The empirical confidence intervals were determined as follows: The distribution of values for a variable was inspected, and the largest value not exceeding more than 2.5% of the distribution was selected as the lower limit of the 95% confidence interval. The smallest value exceeded by 2.5% or less of the distribution was selected as the upper limit of the 95% confidence interval. When there were fewer than 40 data cases in the distribution, the 95% empirical confidence interval (as here defined) was identical to the range of values. When there were 40 or more data cases, the range and empirical confidence interval did not necessarily coincide.

The confidence interval usually reported is a theoretical confidence interval based on the assumption of an underlying normal distribution. The 95% normal distribution confidence interval (95% Normal CI) is given by the following expression:

$$95\% \text{ Normal CI} = \bar{X} \pm 1.96 \text{ (SD)}.$$

Two considerations led us to report empirical confidence intervals rather than normal distribution confidence intervals. First, the normal distribution confidence interval can be easily computed given the mean and standard deviation. In this study, reporting the normal distribution confidence interval is redundant because the mean and standard deviation are reported. In contrast, an empirical confidence interval cannot be determined from distributional characteristics, such as the mean and standard deviation. Rather, an empirical confidence interval must be determined from the actual frequency distribution of the data cases. Second, the normal distribution confidence interval is valid only if the distribution is approximately normal. The distribution for many of the variables presented in Table A-5 are far from normal. It would therefore have been misleading to report normal curve confidence intervals for all variables.

Consider an example in which both types of confidence intervals have been determined for a variable with a nonnormal distribution. Figure A-2 presents the frequency distribution and the overall distribution statistics for the pristane-to-phytane ratio in sediments. Note that the distribution is skewed to the right (Skew. = 2.336) and is more peaked than a normal distribution (Kurt. = 6.869). The skewness and kurtosis values differ from 0 at the 0.01 level of significance, according to critical value tables (Snedecor and Cochran, 1967). Both the empirical and normal distribution confidence intervals have been drawn above the frequency distribution. Note that the two confi-

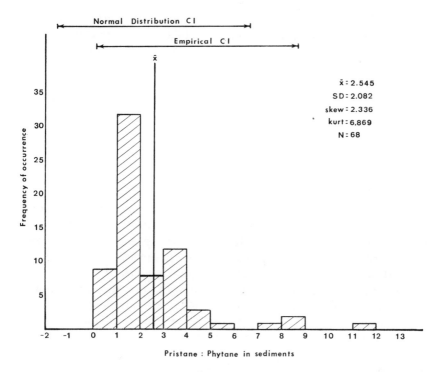

Figure A-2. Example comparison of empirical and normal curve confidence intervals.

dence intervals are quite different. Problems with the normal distribution confidence interval can readily be seen. The lower limit is −1.5, an impossible value for the pristane-to-phytane ratio. The upper limit of 6.6 is exceeded by almost 6% of the data values, and this upper limit probably severely underestimates the true value for the underlying population of values.

While empirical confidence intervals are generally applicable, they should be interpreted with some caution. The empirical confidence interval is most greatly influenced by the extreme values in a distribution. Extreme values are most subject to sampling error, and therefore the empirical confidence interval can be greatly influenced by sampling error. In contrast, the normal distribution confidence interval will be less of a function of sampling error since it is not so dependent upon values that are the extreme and most susceptible to error in a distribution.

All of these considerations about confidence intervals suggest the following recommendations to a user of Table A-5. Inspect the skewness and kurtosis values for a variable. If neither is significantly different from 0, then assume that the distribution is normal. The normal distribution confidence interval should then be calculated and used to best describe that variable. If either skewness or kurtosis is significantly different from 0, then assume that the distribution is not normal. In this case, the empirical confidence interval should best describe that variable.

Analysis of Spatial and Temporal Effects

Each variable in Table A-5 was analyzed according to temporal and spatial variation. The analysis procedures employed were more complicated than one might anticipate. The complexity arose for two types of reasons. First, from a statistical point of view, several aspects of the design of the STOCS study were quite haphazard. The purpose of the study evolved from year to year with corresponding design changes occurring from year to year. Replicate samples (a series of samples taken for each collection period and site combination) were taken inconsistently, thereby precluding use of the most straightforward statistical designs. Missing data further aggravated our problems. Second, time constraints ruled out the use of different analysis approaches for different variables. It was necessary to arrive at an automated system that could uniformly be applied to all variables. Such a uniform approach further sacrificed analytic simplicity.

The temporal effects analyzed were collection period and year, and the spatial effects analyzed were station and transect. For many variables replicate samples were not taken consistently and were therefore scattered over the different sampling sites and times. To allow a uniform approach to all variables, data from replicate samples were averaged to arrive at a single mean data case for each site-period-year combination. These mean values were then analyzed for temporal and spatial variation.

For study elements involving body burdens, desired samples were often not obtained due to failure to catch the species in question. For other study elements (e.g., high molecular-weight hydrocarbons in sediment), the contracted samples involved one set of sites during one collection period but a different set of sites during other collection periods. Thus, for several variables the data were scattered over the range of possible data cases. Even when samples were obtained, it was often the case that particular variables were uncalculable or unmeasurable. For example, variables involving hydrocarbon ratios (e.g., pristane:phytane) were uncalculable if the concentration in the denominator was 0. Trace metal concentrations were sometimes unmeasurable due to detection limit problems.

When data cases were scattered over the possible collection sites and times or when there were missing data for some data cases, analyses for temporal and spatial variation involved unbalanced data (i.e., unequal cell frequencies). Standard analysis of variance (ANOVA) calculation techniques (involving simple comparison of means) are not useful with unbalanced data. When data are unbalanced, all effects (both main effects and interactions) are confounded (Kerlinger and Pedhazur 1973; Rao 1965; Searle 1971). Consider Figure A-3. Numbers within the cells are numbers of data cases. Design 1 involves balanced data (equal cell frequencies), while Design 2 involves unbalanced data (unequal cell frequencies). The standard ANOVA technique for assessing the seasonal effect in such designs involves comparisons of the mean value for season 1 with the mean value for season 2. For Design 1, the season 1 mean involves data cases that are equally divided between Transect I and II, and the same is true for the season 2 mean. Therefore, the difference between the season 1 mean and the season 2 mean is balanced according to transect. In Design 1, effects are unconfounded, and the standard ANOVA procedure of comparing row means and comparing column means is appropriate. For Design 2, the season 1 mean involves data cases from Transect I 20% of the time and data cases from Transect II 80% of the time. The season 2 mean for this design involves equal numbers of Transect I and Transect II data cases. If transect does indeed have an effect, then this differential effect due to the two transects will produce a difference

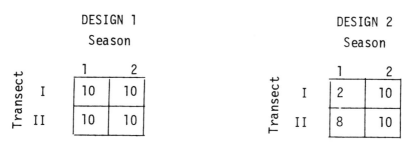

Figure A-3. Cells demonstrating unbalanced data collections that confound calculation of effects.

between the season 1 mean and the season 2 mean, regardless of whether the two seasons have differential effects or not. In Design 2, all the effects (transect, season, and transect by season interaction) are confounded and standard ANOVA procedures are not valid.

Multiple linear regression analysis is the technique suggested for analysis of unbalanced data (Kerlinger and Pedhazur 1973; Rao 1965; Searle 1971). For the study data, multiple linear regression analysis was used to assess the effect of a factor with all other factors in the design covaried (statistically controlled). For example, for a two-way analysis involving transect and season, the transect effect was assessed with the season effect, and the transect by season interaction covaried; the season effect was assessed with the transect effect, and the transect by season interaction covaried; and the transect by season interaction was assessed with the transect effect, and the season effect covaried. All regression analyses were calculated by using the "regression option" of subprogram ANOVA from the *Statistical Package for the Social Sciences* (Nie et al. 1975).

Regression analysis with covaried effects was applied to all the variables reported in Table A-5 whether the data for those variables were balanced or unbalanced. Such a uniform approach to all data was quite satisfactory. For variables with unbalanced data, regression analysis with covaried effects was necessary for meaningful interpretation of results. For variables with balanced data, regression analysis with covaried effects produced exactly the same results and conclusions as standard ANOVA procedures would have (Searle 1971).

For most variables, there was an insufficient number of data cases to attempt a full four-factor design simultaneously incorporating all four effects of interest (transect, station group, collection period, and year). To allow a uniform approach to all variables, a series of two-factor analyses were performed for each variable. Table A-4 presents the two-factor analyses performed for those variables sampled according to the 12-station scheme, for those sampled according to the 25-station scheme, and for those sampled according to the 2-station scheme. For the 12-station scheme, all possible two-factor analyses were performed. For the 25-station scheme, five of the six possible two-factor analyses were performed. The transect by station group analysis was not attempted for the 25-station sampling scheme. A glance at Table A-3 will demonstrate the difficulty in performing a transect-by-station analysis for the 25-station sampling scheme. The transects are haphazardly represented in the first three station depth groups. Note that there is no easy redefinition of these three station groups that would yield groups containing an equal number of representatives from each transect. Given this situation, the results of a

Table A-4. Two-Factor Analyses Strategy Performed for Variables Sampled According to Different Sampling Schemes

Sampling Scheme	Analyses Performed
12-Station	Transect (I–IV) by Station Group (1–3) Transect (I–IV) by Period (1–9) Transect (I–IV) by Year (1975–1977) Station Group (1–3) by Period (1–9) Station Group (1–3) by Year (1975–1977) Period (1–9) by Year (1975–1977)
25-Station	Transect (I–IV) by Period (1–9) Transect (I–IV) by Year (1976–1977) Station Group (1–6) by Period (1–9) Station Group (1–6) by Year (1976–1977) Period (1–9) by Year (1976–1977)
2-Station	Transect (HR–SB) by Period (1–9) Transect (HR–SB) by Year (1976–1977) Period (1–9) by Year (1976–1977)

transect-by-station analysis would have been quite difficult to interpret. For the 2-station sampling scheme, only three two-factor analyses were performed. For the 2-station scheme, there was only one spatial effect (transect). This one spatial effect with the two temporal effects (period and year) produced three possible two-factor analyses.

Note that a significance level of ≤ 0.05 was employed in all analyses for spatial and temporal effects. The following procedure was employed in order to lessen the probability of accepting a chance-produced significant result as a valid result: The overall F ratio for each two-factor analysis was examined. These overall Fs are analogous to the overall between-groups Fs in standard ANOVA; they provide a single test of all effects (main effects and interaction) pooled together. If the overall F for a specific two-factor analysis was not significant (using the $P \leq 0.05$ level), then the entire set of results for that analysis was discarded as chance produced. If the overall F was significant, then significant main effects from that analysis were accepted as valid significant results. In other words, a significant main effect was accepted as valid only if the corresponding overall F was also significant.

The entire set of two-factor analyses for a given variable was then inspected. Only if a given effect (e.g., year) was significant in every two-factor analyses involving that effect was that effect accepted as a clear source of significant variation. For example, consider a variable collected under the 12-station sampling scheme. Six two-factor analyses would be involved in this case, and the year effect would be analyzed in three of the six analyses. If year were found to be significant in each of the three analyses, then year would be accepted as clearly significant. That is, year is significant when period is covaried, when station is covaried, and when transect is covaried. If year were found to be significant in only one or two of the three analyses, then the picture would be unclear. The significant year effects in one or two of the analyses do indicate significant variation, but clear identification of the source of this significant variation is not possible due to confounded effects.

If a main effect was accepted as being clearly significant for a particular variable, then all interactions involving that main effect were inspected for significance. A significant interaction involving a main effect indicated that the main effect may not be general. For example, consider a case in which the main effect of station is significant and the station by transect interaction is also significant. The significant station main effect indicates that stations differ on the average. The significant station by transect interaction indicates that the difference among stations varies for the different transects. It is quite possible that stations are different on Transects I and II but not on Transects III and IV. That is, the station effect may not be general according to transect. Because of such possibilities, significant main effects have been reported in Table A-5 only when there were no significant interactions involving those main effects.

A few comments are necessary concerning these procedures for selection of spatial and temporal effects. For some variables, a limited number of data cases resulted in two-factor designs with empty cells. In these cases it was impossible to evaluate the two-way interaction. Also, for some trace metal body burden variables, data were not available for an entire spatial category (e.g., Transect II or Station Group 3) or any entire temporal category (e.g., spring). In such cases, these categories were omitted from analysis.

In summary, a spatial or temporal result was included in Table A-5 only if the answer to all of the following questions was "yes."

1. Is the overall F significant for every two-factor analysis involving the main effect in question?
2. Is the main effect significant in each of the relevant two-factor analyses?
3. Are the interactions involving the main effect all insignificant?

This procedure for selecting the temporal and spatial results for inclusion in Table A-5 served to limit the reported effects to those which were clear, general, and had the least probability of being chance produced.

Comparison of the Present Base Line Results with Future Monitoring Results

The *t*-test presents a simple method for comparing the present base line results with the results obtained in future monitoring. Given a mean and standard deviation from the present base line results and a mean and standard deviation from future monitoring, a *t* value can be computed as follows:

$$t = \frac{\bar{X}_M - \bar{X}_B}{SE}. \tag{1}$$

In expression (1), $\bar{X}_M$ indicates the mean value for future monitoring, $\bar{X}_B$ indicates the mean base line value, and SE indicates the appropriate standard error. The standard error in expression (1) is the standard error of the difference between the two means and this standard error is given as follows:

$$SE = \left[\left(\frac{SS_B + SS_M}{N_B + N_M - 2}\right)\left(\frac{1}{N_B} + \frac{1}{N_M}\right)\right]^{1/2}. \tag{2}$$

The terms in expression (2) are as follows: SS_B indicates the sum of squares for the base line results, SS_M indicates the sum of squares for the monitoring results, N_B indicates the number of data cases for the base line results, and N_M indicates the number of data cases for the monitoring results. The monitoring sum of squares is defined as follows:

$$SS_M = \sum_{i=1}^{N_M} (\bar{X}_{M_i} - \bar{X}_M)^2.$$

The base line sum of squares can be obtained from the standard deviation (SD) in Table A-5 as follows:

$$SS_B = (SD)(N_B - 1).$$

The *t* value in expression (1) is associated with a number of degrees of freedom equal to $N_B + N_M - 2$. The significance of an obtained *t* can then be determined by reference to a *t*-table with the obtained *t* value and the degree of freedom.

[174] Appendix A

Comparison of base line and monitoring results is quite straightforward when the base line results (Table A-5) do not include significant temporal or spatial variation or both. In this case, the *t*-test can be based upon the overall monitoring results. When the base line results for a variable indicate significant temporal or spatial differences or both, then determining base line and monitoring differences is more complicated. If there is significant base line variation due to collection period (i.e., month or season), then base line and monitoring results comparisons should be performed separately for each collection period involved. If there is significant base line variation due to collection site (i.e., station depth, group, or transect), then separate comparisons should be made for each relevant spatial category.

Determination of the Number of Samples Needed in Future Monitoring

Suppose one wishes to determine if the value for a variable has changed significantly from the base line value. How many monitoring samples (data cases) would be required? This "number of samples" issue falls within the realm of statistical power analysis. The power of a statistical test is the probability that it will yield significant results when a difference actually exists in nature. That is, statistical power is the probability of detecting a population difference in a statistical analysis of samples drawn from that population. For present purposes, we are concerned with the power of the *t*-test for means. The statistical power of the *t*-test is a complex function of the difference between the population means, the level of significance chosen, the sample size (number of data cases), and the measurement error.

Consider the basic formula for the *t*-test:

$$t = \frac{\bar{X}_1 - \bar{X}_2}{SN^{-1/2}}. \tag{3}$$

In expression (3), $\bar{X}_1$ and $\bar{X}_2$ refer to the means of the two groups, S is the standard deviation pooled within groups, and N is the sample size (number of data cases per group). The obtained value of t is significant if it exceeds a critical value. The power of the *t*-test will increase if the calculated value of t increases or if the critical value decreases. Increasing the significance level (i.e., from 0.05 to 0.10) will decrease the critical value. Thus, the power of the *t*-test is a positive function of significance level.

Note in expression (3) that the obtained t value is positively related to the mean differences $(\bar{X}_1 - \bar{X}_2)$. The larger the difference in population means between the two groups, the larger the calculated mean difference will tend to be. Thus, the power of the t-test is a positive function of the population mean difference. The value of S in (3) is a positive function of measurement error. The obtained t in turn is inversely related to S; therefore, the power of the t-test is a negative function of measurement error. The sample size (N) is directly related to the obtained t in expression (3). The power of the t-test is therefore positively related to the sample size.

For present purposes, it will be convenient to define the effect size (EF) as follows:

$$EF = \frac{\bar{X}_1 - \bar{\bar{X}}}{S} - \frac{\bar{X}_2 - \bar{\bar{X}}}{S}. \qquad (4)$$

In expression (4), $\bar{\bar{X}}$ refers to the overall grand mean. The effect size in (4) is a measure of the standardized (normal deviate) difference between the two group means. The convenience of such an effect size is that it is independent of the specific units of measurement. Such an effect size can be used in constructing general tables that are applicable to any variable regardless of units of measurement. Expression (4) can be simplified as follows:

$$EF = \frac{\bar{X}_1 - \bar{X}_2}{S}. \qquad (5)$$

Expression (5) can then be substituted into expression (3) yielding:

$$t = (EF)(N)^{1/2}. \qquad (6)$$

Note in expression (6) that the t-value is a positive function of the effect size. Thus, the power of the t-test is a positive function of effect size.

Sufficient background has been provided to consider now a table of sample sizes based upon statistical power (e.g., Table 2.4.1 in Cohen, 1969, pp. 52–53). Such a table presents the sample sizes required to obtain a significant difference with a given probability (power) for a given effect size (EF). To use this table, one must choose a desired effect size, power, and level of significance. The level of significance is the probability of finding a difference by chance when one does not exist. Selection of significance level is straightforward and needs no discussion. Selection of effect size and power are more difficult and need to be considered in detail.

The effect size of interest in this appendix is based upon a difference between known base line results and the unknown results of future monitoring. For this case, expression (5) can be rewritten as follows:

$$EF = \left| \frac{\bar{X}_M - \bar{X}_B}{S_B} \right|. \tag{7}$$

In expression (7), $\bar{X}_M$ and $\bar{X}_B$ are the monitoring and base line means, while S_B is the base line standard deviation. Note that before monitoring S_B is the best guess of the value of the standard deviation pooled within groups that is calculable after monitoring. Before monitoring, $\bar{X}_M$ is the only unknown value on the right side of expression (7). If one can select a mean for monitoring ($\bar{X}_M$) that is a critical value to detect, then this value can be substituted into (7) thereby yielding the desired effect size for entry into a table to determine sample size. For example, the $\bar{X}_M$ chosen may be the critical value of a trace metal beyond which this metal is toxic in the aquatic environment. Another choice of a critical value for the $\bar{X}_M$ may be one or two standard deviations below the $\bar{X}_B$ to use, for example, for benthos infaunal density or demersal fish density. One must then specify the level of confidence (power) for detecting a value as extreme as the critical value of $\bar{X}_M$. This level of confidence is the probability of finding a difference when one exists and usually is greater than 0.80. Entry into a sample size table with this confidence level (power) and effect size will then yield the required sample size (number of data cases) in the body of the table.

Therefore, when selecting a sample size, the rule of thumb is to have a significance level as low as possible and a power as high as possible. The meeting of these two criteria, however, depends upon the economics of obtaining the information, since by meeting the criteria you will dramatically increase your sample size.

Unfortunately, one difficulty remains in using a sample size value obtained from a table. The sample size values in this table are for a *t*-test involving two groups of equal size. Construction of sample size tables allowing unequal group sizes involves too much additional complexity. A sample size table can be adapted, however, for use with unequal group sizes in the manner suggested by Cohen (1969). For unequal-sized groups, the same size values should equal the harmonic mean of the two group sizes. The harmonic mean ($\bar{N}_h$) of the two group sizes (N_1 and N_2) is given by

$$\bar{N}_h = \frac{2 N_1 N_2}{N_1 + N_2}.$$

For our case involving base line and monitoring results, this expression becomes

$$\bar{N}_h = \frac{2N_B N_M}{N_B + N_M}. \tag{8}$$

Now $\bar{N}_h$ is the value from the sample size table and N_B is the base line sample size, a known value. Therefore, N_M is the only unknown in (8), and this expression can be solved for N_M with the following result:

$$N_M = \frac{\bar{N}_h N_B}{2N_B - \bar{N}_h}. \tag{9}$$

Expression (9) can then be used to estimate the required number of monitoring samples, given the value of ($\bar{N}_h$) and the known number of base line data cases (N_B). One caution must be considered concerning the use of expression (9). Note that the calculated value of N_M will be negative if the quantity, $2N_B - \bar{N}_h$, is less than zero. In this case, there is no possible value for N_M, and there is no way to detect the chosen effect size with the desired power. The only solution is to decrease the effect size of interest or decrease the power desired or both. Such difficulties arise because the number of base line data cases is fixed, and this number imposes limits on the achievable level of power for a given effect size.

In conclusion, statistical power analysis affords a method for estimating the number of samples needed in future monitoring. While the methods are complex and require the user to make subjective judgments about effect size and confidence, they do provide guidelines to assist in monitoring planning. No viable alternative exists.

Notes to Appendix A

1. Text of appendix prepared by Robert C. Godbout.
2. The actual analyses performed and the rationale for these analyses are presented in the methodology section of this appendix.
3. These station schemes did not begin until 1976; consequently, these schemes only cover the 1976 and 1977 sampling years.
4. "Replicate samples" here refers to different samples taken at the same site, collection period, and year.

Table A-5. Distributional Characteristics for Ecosystem Variables of the South Texas Shelf[a]

Variable	Mean	SD	Skew.	Kurt.	95% Empir. CI		N
Pelagic nonliving characteristics							
HYDROGRAPHY (1976–1977)							
Secchi depth (m)	15.93	9.59	0.47	−0.46	2.00–	36.00	108
Station Group 1	6.83	5.10	1.49	1.77	2.00–	21.00	24
Station Group 2	16.13	7.83	0.85	−0.03	5.00–	34.00	24
Station Group 3	22.71	7.39	0.20	0.24	9.00–	39.00	24
1976	18.42	9.24	0.05	−1.03	5.00–	35.00	36
1977	12.03	8.60	1.06	1.28	2.00–	39.00	36
NUTRIENTS (surface; 1976–1977)							
Silicate (μM/l)	2.32	1.68	1.66	3.46	0.16–	6.80	108
Station Group 1	2.47	1.38	0.68	0.59	0.16–	6.00	24
Station Group 2	2.07	1.25	0.92	−0.07	0.70–	4.80	24
Station Group 3	1.43	0.66	0.32	−0.59	0.40–	2.80	24
Phosphate (μM/l)	0.19	0.16	0.93	−0.15	0.01–	0.56	105
Nitrate (μM/l)	0.29	0.51	4.98	29.94	0.05–	1.90	108
Dissolved oxygen (ml/l)	5.10	0.53	1.15	1.33	4.26–	6.36	108
LOW MOLECULAR-WEIGHT HYDROCARBONS (surface, dissolved; 1975–1977)							
Methane (nl/l)	89.47	64.96	4.33	25.30	41.00–	275.00	144
Winter	106.86	77.73	2.12	4.27	40.00–	377.00	36
Spring	73.14	41.63	3.03	11.28	37.00–	260.00	36
Fall	81.28	36.20	2.70	9.86	45.00–	240.00	36
1975	78.11	36.17	2.83	10.93	37.00–	240.00	36
1976	68.42	24.79	1.89	4.08	41.00–	157.00	36
1977	114.75	81.07	1.77	2.69	44.0 –	377.0	36
Ethene (nl/l)	7.08	7.46	3.64	17.99	1.5 –	25.3	144
Winter	3.33	1.65	0.67	0.01	0.2 –	7.5	36

Overall Base Line Results [179]

Fall	9.50	5.50	0.96	0.46	2.9 –	25.0	36
1975	12.02	12.06	2.18	5.53	1.5 –	58.3	36
1976	7.27	4.99	0.77	−0.27	0.2 –	19.1	36
1977	4.15	1.72	1.64	3.51	1.9 –	10.0	36
Ethane (nl/l)							
Winter	0.56	0.51	2.56	8.92	0.1 –	1.8	130
Spring	0.95	0.74	1.54	2.83	0.1 –	3.5	35
Fall	0.34	0.07	0.73	1.04	0.2 –	0.5	23
1975	0.39	0.36	1.91	3.59	0.1 –	1.6	36
1976	1.13	0.90	0.52	0.35	0.1 –	3.5	24
1977	0.35	0.14	1.07	1.41	0.1 –	0.7	35
	0.44	0.20	0.52	−0.30	0.1 –	0.9	35
Propene (nl/l)							
Station Group 1	1.43	0.88	2.02	5.77	0.4 –	4.3	141
Station Group 2	1.74	1.15	1.99	3.75	0.5 –	5.7	36
Station Group 3	1.41	0.80	1.47	3.26	0.5 –	4.2	35
	1.31	0.85	1.85	5.72	0.2 –	4.6	34
Winter	0.83	0.41	0.84	−0.04	0.2 –	1.8	35
Spring	1.87	1.06	2.22	5.06	0.9 –	5.7	36
Fall	1.77	0.91	1.66	2.45	0.8 –	4.4	34
1975	2.14	1.39	0.74	−0.05	0.2 –	5.7	33
1976	1.35	0.42	−0.71	−0.46	0.4 –	1.9	36
1977	1.03	0.38	0.24	−0.04	0.4 –	2.0	36
Propane (nl/l)							
Winter	0.72	0.80	2.90	8.80	0.1 –	3.2	138
Spring	0.83	0.73	2.97	11.22	0.3 –	4.1	35
Fall	0.47	0.26	1.52	3.07	0.1 –	1.3	32
1975	1.14	1.26	1.38	0.73	0.2 –	4.7	35
1976	1.73	1.23	0.74	−0.07	0.1 –	4.7	31
1977	0.42	0.12	1.32	3.27	0.2 –	0.8	36
	0.43	0.13	0.24	0.05	0.2 –	0.7	35
NUTRIENTS (half photic zone: 1976–1977)							
Silicate (µM/l)	2.18	1.60	1.71	3.36	0.15–	6.80	108
Station Group 1	2.50	1.42	0.73	0.74	0.15–	6.20	24

[180] Appendix A

Table A-5—*Continued*

Variable	Mean	SD	Skew.	Kurt.	95% Empir. CI	N
Station Group 2	1.85	0.85	0.32	−0.96	0.70– 3.40	24
Station Group 3	1.27	0.58	0.90	0.12	0.50– 2.60	24
Winter	1.85	1.08	0.81	0.30	0.15– 4.50	24
Spring	1.37	0.59	0.66	−0.94	0.80– 2.50	24
Fall	2.40	1.33	0.99	1.74	0.50– 6.20	24
Phosphate (μM/l)	0.17	0.19	2.32	8.03	0.01– 0.67	107
Nitrate (μM/l)	0.24	0.46	5.50	34.46	0.05– 1.80	108
Dissolved oxygen (ml/l)	5.10	0.46	1.47	2.82	4.47– 6.35	96
LOW MOLECULAR-WEIGHT HYDROCARBONS (half photic zone, dissolved; 1975–1977)						
Methane (nl/l)	155.56	473.90	8.44	75.52	42.00– 393.00	114
Ethene (nl/l)	6.16	4.86	2.46	7.40	1.70– 22.20	107
Winter	3.65	2.09	2.74	10.02	1.3 – 11.8	24
Spring	7.10	4.53	1.47	2.17	1.7 – 22.2	36
Fall	8.24	7.23	1.95	3.38	2.3 – 30.0	24
1975	8.98	6.85	1.43	1.75	1.3 – 30.0	36
1976	5.86	2.90	0.42	−1.15	1.7 – 11.0	12
1977	4.10	1.93	1.94	4.93	1.9 – 11.0	36
Ethane (nl/l)	0.72	0.61	1.43	1.70	0.1 – 2.2	80
Winter	1.17	0.62	1.04	1.50	0.5 – 3.0	24
Spring	0.44	0.18	0.99	0.00	0.2 – 0.8	12
Fall	0.55	0.67	1.79	2.00	0.1 – 2.2	24
Propene (nl/l)	1.30	0.89	1.80	3.79	0.2 – 3.7	89
Propane (nl/l)	0.98	1.08	1.89	2.72	0.01– 4.0	88
Winter	1.08	0.93	2.81	9.96	0.4 – 4.7	24
Spring	0.49	0.31	0.88	1.35	0.01– 1.3	21
Fall	1.77	1.51	0.31	−1.66	0.2 – 4.2	24

(10 m depth, dissolved; 1975–1977)							
Total hydrocarbons (µg/g)	0.22			72.75		0.91	125
($n-C_{14}$ to $n-C_{32}$)							
Pristane:phytane							
Winter	2.73	4.34	5.33	31.10	0.12–	7.17	40
	10.04	12.18	1.81	3.27	2.00–	28.00	4
Spring	2.12	1.50	1.73	4.50	0.12–	6.75	18
Fall	1.72	1.00	0.10	–0.97	0.31–	3.50	14
HIGH MOLECULAR-WEIGHT HYDROCARBONS							
(10 m depth, particulate; 1975–1977)							
Total hydrocarbons (µg/g)	0.18	0.36	3.34	10.80	0.00–	1.61	103
($n-C_{14}$ to $n-C_{32}$)							
Phytane:$n-C_{18}$	0.83	2.21	5.75	34.88	0.03–	2.41	40
1975	0.08	0.02	1.72	0.00	0.07–	0.10	3
1976	0.18	0.13	1.26	0.54	0.03–	0.45	17
1977	0.83	0.67	1.26	1.04	0.09–	2.41	16
SUM MID (relative percent)							
($n-C_{19}$ to $n-C_{24}$)	22.91	21.71	1.30	0.80	0.00–	75.80	101
Winter	32.80	19.60	0.02	–1.14	0.00–	65.54	24
Spring	19.94	22.70	2.19	4.25	0.00–	88.38	26
Fall	14.03	13.70	2.80	10.21	0.00–	73.33	35
NUTRIENTS (bottom; 1976–1977)							
Silicate (µM/l)							
Station Group 1	3.70	2.06	0.61	–0.23	0.40–	7.60	90
Station Group 2	2.24	1.58	0.63	–0.54	0.30–	4.63	6
Station Group 3	3.63	2.10	0.30	–1.51	0.80–	6.70	18
Station Group 4	3.57	2.28	1.42	2.33	0.80–	9.98	18
Station Group 5	3.66	2.25	0.98	1.26	1.33–	7.54	6
Station Group 6	2.64	1.17	–0.26	–0.59	0.40–	4.20	12
	5.48	'72	0.15	–1.76	3.30–	7.80	12
Transect I	4.00	26	–0.09	–1.07	0.30–	7.60	18
Transect II	4.60	45	0.60	–0.58	1.60–	9.98	18
Transect III	3.49	.77	0.68	0.17	0.97–	7.54	18

Overall Base Line Results [181]

[182] Appendix A

Table A-5—*Continued*

Variable	Mean	SD	Skew.	Kurt.	95% Empir. CI		N
Transect IV	2.50	1.30	0.78	1.13	0.40–	5.81	18
Phosphate (μM/l)	0.48	0.62	4.40	26.30	0.03–	1.88	89
Station Group 1	0.28	0.14	−0.11	−2.52	0.11–	0.43	6
Station Group 2	0.38	0.30	2.17	6.09	0.03–	1.34	17
Station Group 3	0.25	0.17	1.21	0.96	0.03–	0.67	18
Station Group 4	0.23	0.12	0.57	−2.01	0.10–	0.38	6
Station Group 5	0.30	0.31	2.36	6.55	0.02–	1.19	12
Station Group 6	1.16	1.23	2.58	7.53	0.17–	4.74	12
Nitrate (μM/l)	3.64	5.48	1.75	2.01	0.10–	17.10	81
Dissolved oxygen (ml/l)	4.43	0.95	−0.50	−0.78	2.58–	5.86	90
Station Group 1	4.53	1.35	−0.99	−0.18	2.32–	5.86	6
Station Group 2	4.75	0.82	−0.24	−0.66	3.06–	6.00	18
Station Group 3	4.83	0.57	−0.14	−0.37	3.70–	5.83	18
Station Group 4	4.71	0.52	0.12	−0.61	3.98–	5.43	6
Station Group 5	4.21	0.91	−0.52	−0.81	2.58–	5.37	12
Station Group 6	3.58	0.96	0.95	−0.07	2.66–	5.61	12
Winter	4.98	1.00	−1.35	0.60	2.80–	6.00	24
Spring	4.24	0.82	−0.84	0.01	2.32–	5.43	24
Fall	4.17	0.75	−0.71	−0.02	2.58–	5.47	24
1976	4.16	1.01	−0.36	−1.25	2.32–	5.71	36
1977	4.76	0.73	−0.05	−1.00	3.37–	6.00	36

LOW MOLECULAR-WEIGHT HYDROCARBONS
(bottom, dissolved; 1975–1977)

Variable	Mean	SD	Skew.	Kurt.	95% Empir. CI		N
Methane (nl/l)	190.40	150.89	1.87	3.85	45.0 –	589.0	88
Ethene (nl/l)	3.25	2.63	2.98	13.63	0.1 –	10.5	89
Station Group 1	4.36	2.33	1.05	−0.73	2.4 –	8.1	6
Station Group 2	4.19	2.29	2.41	7.01	1.9 –	11.8	18
Station Group 3	4.49	4.11	2.77	8.53	0.7 –	18.6	18
Station Group 4	3.61	1.11	1.06	2.86	2.2 –	5.6	6

Overall Base Line Results

	Mean	SD	Skewness	Kurtosis	Min	Max	N
Winter	2.12	1.04	0.08	−0.48	0.1−	4.0	24
Spring	3.21	1.29	0.08	0.79	0.7−	6.4	24
Fall	5.01	4.08	1.89	4.38	0.8−	18.6	24
Ethane (nl/l)	0.54	0.25	1.26	3.11	0.1−	1.0	89
Propene (nl/l)							
Station Group 1	0.65	0.35	0.84	−0.15	0.2−	1.5	88
Station Group 2	1.16	0.38	−0.93	−0.28	0.6−	1.5	6
Station Group 3	1.01	0.30	−0.17	−0.78	0.5−	1.5	18
Station Group 4	0.62	0.18	0.46	0.26	0.3−	1.0	17
Station Group 5	0.51	0.13	−0.48	−0.97	0.3−	0.7	6
Station Group 6	0.48	0.13	1.61	3.09	0.3−	0.8	12
Station Group 6	0.32	0.06	−0.64	−0.11	0.2−	0.4	12
Propane (nl/l)	0.52	0.17	1.40	2.71	0.3−	1.0	88

Pelagic living characteristics

PHYTOPLANKTON (surface; 1976–1977)

	Mean	SD	Skewness	Kurtosis	Min	Max	N
Phytoplankton species (species/liter)	28.94	15.94	0.69	0.04	5.00−	67.00	107
Phytoplankton density (cells/liter)	83,613.38	261,934.61	6.63	50.57	400.00−	963,392.00	107
Chaetoceros spp. (cells/liter)							
Transect I	5,283.28	20,848.99	7.52	63.51	0.00−	30,200.00	107
Transect II	7,147.24	10,107.56	1.48	1.27	0.00−	30,200.00	17
Transect III	3,052.33	6,590.06	3.10	10.51	0.00−	26,720.00	18
Transect IV	1,157.39	2,119.01	2.63	6.76	0.00−	8,017.00	18
Transect IV	1,509.11	3,583.39	2.73	6.62	0.00−	127,400.00	18
Nanno ^{14}C (mgC/m^3/hr)	3.20	3.33	1.25	1.28	0.00−	11.71	51
Net ^{14}C (mgC/m^3/hr)	2.38	3.52	1.66	1.80	0.00−	11.89	51
Total ^{14}C (mgC/m^3/hr)	4.99	5.37	1.21	1.09	0.00−	19.06	53
Nannochlorophyll (μg/l)	0.45	0.35	1.84	4.41	0.00−	1.59	107
Net chlorophyll (μg/l)	0.19	0.42	3.32	11.82	0.00−	1.93	107

[184] Appendix A

Table A-5—*Continued*

Variable	Mean	SD	Skew.	Kurt.	95% Empir. CI		N
Station Group 1	0.33	0.37	1.58	1.81	0.00–	1.26	24
Station Group 2	0.05	0.08	1.87	3.60	0.00–	0.28	24
Station Group 3	0.04	0.12	3.10	8.44	0.00–	0.41	23
Total chlorophyll (µg/l)	0.64	0.66	1.95	3.43	0.00–	2.77	107
Winter	0.80	0.57	1.39	1.06	0.25–	2.20	24
Spring	0.43	0.35	1.36	2.31	0.00–	2.00	24
Fall	0.58	0.58	1.82	2.85	0.15–	2.31	23
Nanno phaeophytin (µg/l)	1.20	0.34	−2.72	7.69	0.00–	1.47	107
Station Group 1	1.35	0.15	2.55	9.29	1.18–	1.93	24
Station Group 2	1.21	0.28	−3.80	16.63	0.00–	1.42	24
Station Group 3	1.13	0.37	−2.76	7.09	0.00–	1.44	23
Winter	1.31	0.09	−0.80	0.81	1.06–	1.44	24
Spring	1.11	0.44	−2.27	3.77	0.00–	1.45	24
Fall	1.26	0.19	2.14	6.80	1.06–	1.93	23
1976	1.13	0.36	−2.72	6.72	0.00–	1.44	36
1977	1.33	0.14	2.39	10.48	1.11–	1.93	35
Net phaeophytin (µg/l)	0.66	0.75	0.29	−1.85	0.00–	1.67	107
Station Group 1	1.20	0.65	−1.27	0.04	0.00–	2.00	24
Station Group 2	0.66	0.74	0.25	−2.02	0.00–	1.67	24
Station Group 3	0.20	0.54	2.36	3.89	0.00–	1.60	23
1976	0.37	0.66	1.35	0.05	0.00–	2.00	36
1977	1.03	0.72	−0.76	−1.41	0.00–	1.75	35
Total phaeophytin (µg/l)	1.22	0.35	−2.67	7.27	0.00–	1.52	107
1976	1.15	0.37	−2.61	6.29	0.00–	1.47	36
1977	1.36	0.13	1.15	4.27	1.11–	1.85	35
PHYTOPLANKTON (half photic zone; 1976–1977)							
Phytoplankton species (species/liter)							
1976	28.18	14.82	0.71	0.37	3.00–	68.00	108
1976	34.67	13.81	0.70	0.18	13.00–	73.00	36

Overall Base Line Results [185]

(cells/liter)							
Winter	71,671.67	177,591.70	4.44	20.94	60.00–	478,729.00	108
Spring	73,766.33	102,835.75	2.95	10.40	5,660.00–	478,729.00	24
	19,573.38	25,451.93	3.42	13.46	2,260.00–	125,105.00	24
Fall	30,797.75	82,587.35	4.50	21.08	60.00–	407,685.00	24
Nannochlorophyll (μg/l)	0.46	0.38	2.07	6.52	0.00–	1.51	108
Net chlorophyll (μg/l)	0.17	0.36	3.15	10.62	0.00–	1.46	108
Total chlorophyll (μg/l)	0.63	0.67	2.09	4.45	0.00–	2.75	108
Nannophaeophytin (μg/l)	1.16	0.38	−2.54	5.17	0.00–	1.46	108
Net phaeophytin (μg/l)	0.65	0.77	0.42	−1.63	0.00–	1.80	108
Station Group 1	1.14	0.68	−1.15	−0.62	0.00–	1.67	24
Station Group 2	0.54	0.80	1.04	−0.40	0.00–	2.50	24
Station Group 3	0.33	0.66	1.67	1.11	0.00–	2.00	24
1976	0.33	0.63	1.42	0.07	0.00–	1.60	36
1977	1.00	0.79	−0.34	−1.47	0.00–	2.50	36
Total phaeophytin (μg/l)	1.20	0.36	−2.76	6.87	0.00–	1.51	108
PHYTOPLANKTON (bottom; 1976–1977)							
Total chlorophyll (μg/l)	0.75	0.65	1.65	403	0.00–	2.18	90
Station Group 1	1.29	0.73	−0.52	−1.38	0.23–	2.12	6
Station Group 2	1.21	0.59	0.88	0.78	0.24–	2.63	18
Station Group 3	0.70	0.35	0.33	−0.86	0.11–	1.32	18
Station Group 4	0.62	0.24	0.84	0.74	0.38–	1.02	6
Station Group 5	0.32	0.17	−0.10	0.12	0.00–	0.58	12
Station Group 6	0.18	0.19	0.75	0.08	0.00–	0.59	12
Total phaeophytin (μg/l)	1.19	0.39	−2.53	5.38	0.00–	1.50	90
ZOOPLANKTON (1975–1977)							
Copepod species number (species/m³)	41.03	20.47	0.57	−0.85	15.00–	83.00	144
Copepod total density (individuals/m³)	535.94	470.40	3.28	15.62	91.30–	1,768.10	144
Station Group 1	816.54	742.39	2.18	5.58	47.00–	3,594.10	36

Table A-5—Continued

Variable	Mean	SD	Skew.	Kurt.	95% Empir. CI	N
Station Group 2	478.14	273.77	0.69	0.01	56.40– 1,230.10	36
Station Group 3	331.37	176.03	1.63	4.07	107.40– 977.60	36
Ichthyoplankton density (larvae/1000 m³)	2,091.28	1,741.02	1.33	1.92	75.00– 6,209.00	144
Winter	1,000.29	1,044.07	1.90	5.20	32.00– 5,056.50	36
Spring	2,254.28	1,694.90	0.99	−0.25	329.50– 5,975.00	36
Fall	2,829.04	1,804.82	1.34	2.82	361.00– 8,932.00	36
Total zooplankton biomass (g/m³)	21.84	16.83	1.64	2.33	3.30– 70.60	144
Farranula gracilis (individuals/m³)	17.38	34.78	3.84	18.84	0.00– 105.60	144
Station Group 1	4.94	21.49	5.87	34.87	0.00– 129.40	36
Station Group 2	18.59	27.14	1.63	1.59	0.00– 97.90	36
Station Group 3	22.52	42.78	4.48	23.32	0.00– 248.70	36
Winter	2.54	4.53	2.46	5.98	0.00– 19.85	36
Spring	8.61	15.08	3.42	13.74	0.00– 79.15	36
Fall	34.91	48.44	2.79	10.34	0.00– 248.70	36
Clausocalanus jobei (individuals/m³)	16.28	42.77	5.59	37.44	0.00– 85.50	144
1975	5.81	15.63	3.59	13.30	0.00– 76.05	36
1976	11.92	22.18	2.19	3.68	0.00– 75.95	36
1977	40.09	76.31	3.13	10.16	0.00– 342.55	36
Nannocalanus minor (individuals/m³)	3.51	4.90	1.90	3.86	0.00– 16.75	144
Station Group 1	1.16	2.67	3.00	9.49	0.00– 12.30	36
Station Group 2	3.75	4.46	1.61	2.28	0.00– 17.05	36
Station Group 3	5.73	6.40	1.54	2.08	0.00– 26.15	36
Winter	5.39	5.55	1.15	0.56	0.00– 20.60	36
Spring	2.37	3.12	1.76	2.72	0.00– 12.30	36

Overall Base Line Results [187]

(individuals/m³)							
Station Group 1	15.45	21.11	2.20	6.51	0.00–	65.65	144
Station Group 2	1.88	3.88	3.31	13.23	0.00–	19.90	36
Station Group 3	14.48	19.38	2.74	10.49	0.00–	100.75	36
Station Group 3	32.48	25.70	1.62	3.81	1.55–	121.90	36
Paracalanus aculeatus							
(individuals/m³)							
Winter	23.19	31.78	3.32	14.79	0.00–	122.65	144
Spring	16.47	14.03	1.18	0.89	0.65–	57.85	36
Spring	11.79	14.07	2.74	9.96	0.50–	73.35	36
Fall	38.88	47.15	2.54	7.43	0.00–	219.05	36
Temora turbinata							
(individuals/m³)							
Station Group 1	34.59	86.90	5.18	31.26	0.00–	316.00	144
Station Group 2	83.48	145.84	3.11	10.11	0.00–	689.90	36
Station Group 3	20.33	34.74	3.35	12.17	0.00–	172.45	36
Station Group 3	5.93	10.71	2.82	8.21	0.00–	47.40	36
Paracalanus indicus							
(individuals/m³)							
Station Group 1	81.44	157.24	3.32	12.92	0.00–	634.50	144
Station Group 2	169.81	215.82	1.29	0.39	0.00–	688.45	36
Station Group 3	55.83	54.07	0.91	–0.06	0.00–	192.60	36
Station Group 3	14.60	34.37	5.03	27.67	0.15–	203.35	36
Winter	124.55	149.73	2.09	4.99	0.80–	674.25	36
Spring	85.54	160.64	2.67	6.60	0.00–	688.45	36
Fall	30.14	105.48	5.68	33.28	0.00–	634.50	36
Paracalanus quasimodo							
(individuals/m³)							
Station Group 1	77.01	161.39	5.97	47.32	0.00–	486.45	144
Station Group 2	137.13	139.48	1.20	0.76	0.00–	490.35	36
Station Group 3	48.51	43.18	0.98	0.14	1.05–	160.25	36
Station Group 3	11.64	20.84	3.58	13.15	0.00–	102.30	36
Centropages velificatus							
(individuals/m³)							
Station Group 1	13.83	34.66	4.63	22.85	0.00–	150.65	144
Station Group 2	27.19	43.98	2.87	8.68	0.00–	206.05	36
Station Group 3	17.94	49.21	3.88	14.28	0.00–	220.50	36
Station Group 3	2.89	5.73	3.11	10.11	0.00–	27.30	36

Table A-5—Continued

Variable	Mean	SD	Skew.	Kurt.	95% Empir. CI		N
Total calanoids (individuals/m³)	607.18	667.47	3.60	18.24	61.20–	2,860.20	144
Station Group 1	1,055.78	1,050.89	2.42	6.60	55.85–	5,211.85	36
Station Group 2	534.76	343.48	0.90	0.51	61.20–	1,411.05	36
Station Group 3	258.78	189.90	2.43	7.72	67.75–	1,040.20	36
Larvacea (individuals/m³)	49.22	53.45	2.16	5.90	1.75–	196.30	144
1975	23.25	23.45	2.34	7.52	1.20–	119.65	36
1976	51.49	55.16	2.29	7.11	2.25–	277.35	36
1977	45.69	40.83	1.30	1.63	2.05–	176.35	36
Total Cladocera (individuals/m³)	28.84	70.35	4.30	23.51	0.00–	232.45	144
Station Group 1	56.00	113.48	3.00	10.20	0.00–	545.20	36
Station Group 2	27.58	60.33	2.57	5.94	0.00–	249.85	36
Station Group 3	5.45	8.94	2.44	6.15	0.00–	37.60	36
HIGH MOLECULAR-WEIGHT HYDROCARBON BODY BURDENS (zooplankton; 1975–1977)							
Total hydrocarbons (µg/g) (n–C_{14} to n–C_{32})	134.94	422.68	9.21	93.12	0.58–	436.59	124
Pristane:phytane	351.40	753.79	4.51	24.16	2.66–	101.69	84
Pristane:n–C_{17}	7.80	14.32	4.03	17.44	0.08–	75.12	119
Phytane:n–C_{18}	1.06	5.02	8.03	67.94	0.01–	3.10	85
(Pristane + phytane):n-alkanes	2.30	5.18	6.86	56.68	0.05–	9.91	121
SUM LOW (relative percent) (n–C_{14} to n–C_{18})	56.97	26.10	–0.15	–0.68	0.00–	100.00	123
SUM MID (relative percent) (n–C_{19} to n–C_{24})	26.49	18.67	0.85	1.08	0.00–	61.36	123

					Range		N
$(n-C_{25}$ to $n-C_{32})$	16.55	20.94	1.37	1.46	0.00–	66.14	123
Average OEP (odd-even preference)							
$(n-C_{14}$ to $n-C_{32})$	5.37	8.41	3.35	12.10	0.69–	36.74	103
TRACE METAL BODY BURDENS							
(zooplankton; 1976–1977)							
Iron (ppm dry weight)	2,750.45	3,716.53	1.99	3.68	27.3–	13,333.3	72
Calcium (ppm dry weight)	38,044.09	20,582.30	1.47	3.24	8,200.0–	90,000.0	62
Vanadium (ppm dry weight)	18.08	19.99	2.68	9.90	1.3–	70.0	105
Zinc (ppm dry weight)	157.06	183.80	5.48	34.30	32.00–	500.00	141
Aluminum (ppm dry weight)	5,280.86	7,319.55	1.90	3.00	90.00–	29,000.00	67
Lead (ppm dry weight)	12.60	19.73	4.61	27.67	0.60–	64.00	136
Nickel (ppm dry weight)	7.03	5.35	2.85	12.11	2.00–	20.00	144
Copper (ppm dry weight)	26.34	95.09	11.12	128.73	5.50–	75.00	143
Chromium (ppm dry weight)	3.99	3.08	0.98	0.50	0.10–	11.10	140
Cadmium (ppm dry weight)	3.32	1.54	0.26	−0.62	0.80–	6.50	144
Benthic nonliving characteristics							
SEDIMENT TEXTURE (1976–1977)							
Sediment mean grain size							
(ϕ units)							
Station Group 1	7.70	1.72	−0.75	−0.47	3.69–	9.79	186
Station Group 2	4.85	0.96	0.27	−1.05	3.44–	6.46	24
Station Group 3	6.79	1.63	−0.49	−1.55	4.38–	8.69	18
Station Group 4	7.65	1.07	−0.25	−0.48	5.40–	9.23	36
Station Group 5	8.12	1.44	−1.02	−0.17	4.97–	10.08	24
Station Group 6	8.20	1.56	−0.82	−0.68	5.01–	9.85	24
Transect I	9.47	0.26	−0.20	−0.74	8.98–	9.95	24
Transect II	7.50	1.59	−0.23	−1.08	4.55–	10.08	36
Transect III	8.32	0.94	0.04	−1.27	6.63–	9.85	36
Transect III	8.23	1.87	−1.62	1.18	3.69–	9.83	36
Transect IV	6.37	2.00	0.54	−0.98	3.44–	9.78	42

Table A-5—Continued

Variable	Mean	SD	Skew.	Kurt.	95% Empir. CI	N
Sediment grain size standard deviation (ϕ units)						
Station Group 1	3.26	0.43	−0.04	1.69	2.47– 4.22	186
Station Group 2	3.09	0.61	−1.01	1.05	1.38– 3.78	24
Station Group 3	3.44	0.24	1.26	1.02	3.14– 4.00	18
Station Group 4	3.49	0.33	1.04	0.42	2.94– 4.28	36
Station Group 5	3.37	0.49	0.80	−0.50	2.61– 4.34	24
Station Group 6	3.28	0.49	0.72	−0.60	2.71– 4.38	24
Transect I	2.89	0.17	0.04	−0.88	2.58– 3.21	24
Transect II	3.34	0.28	−0.59	−0.05	2.61– 3.78	36
Transect III	3.22	0.25	−0.68	−0.68	2.71– 3.59	36
Transect IV	2.93	0.39	−2.03	6.33	1.38– 3.40	36
Transect V	3.54	0.58	−0.27	−1.47	2.58– 4.34	42
Percent sand	22.86	24.44	1.17	0.05	1.03– 81.78	186
Station Group 1	67.1	14.9	−0.5	−1.1	39.7 – 87.3	24
Station Group 2	32.3	26.5	0.7	−1.5	4.5 – 74.2	18
Station Group 3	20.5	14.6	0.8	−0.3	2.5 – 55.3	36
Station Group 4	18.1	20.6	1.2	−0.3	2.2 – 61.6	24
Station Group 5	17.4	20.9	1.2	−0.1	1.0 – 60.9	24
Station Group 6	3.5	2.5	0.7	−0.6	0.4 – 9.3	24
Transect I	27.2	21.8	0.7	−0.9	2.2 – 74.4	36
Transect II	11.5	8.0	0.5	−0.6	1.0 – 30.6	36
Transect III	17.0	28.8	1.8	1.6	0.9 – 87.3	36
Transect IV	44.3	26.8	−0.5	−1.1	0.4 – 81.8	42
Percent silt	30.37	10.66	−0.58	−0.54	8.64– 47.55	186
Station Group 1	14.7	6.9	0.7	−0.8	6.9 – 27.8	24
Station Group 2	30.7	14.3	−0.6	−1.6	9.6 – 49.0	18
Station Group 3	34.4	8.7	−0.4	−0.4	14.6 – 48.8	36
Station Group 4	29.8	10.2	−0.6	−1.2	11.9 – 43.1	24
Station Group 5	29.9	8.7	−1.0	−0.3	13.3 – 40.2	24
Station Group 6	29.6	1.8	0.6	−0.0	26.3 – 33.7	24

Overall Base Line Results [191]

	Mean	SD	Skewness	Kurtosis	Range	N
Transect II	37.1	6.2	0.3	−0.7	26.3 – 49.0	36
Transect III	31.5	10.5	−1.2	0.5	8.6 – 44.3	36
Transect IV	18.6	8.0	0.3	−1.3	6.9 – 31.4	42
Percent clay	46.77	17.42	−0.44	−0.69	8.87 – 71.61	186
Station Group 1	18.2	8.8	0.1	−1.2	2.9 – 32.9	24
Station Group 2	37.5	13.6	−0.3	−1.4	15.3 – 55.1	18
Station Group 3	45.1	10.6	0.5	−1.1	30.1 – 62.8	36
Station Group 4	52.1	10.2	−0.6	−1.2	26.5 – 74.8	24
Station Group 5	52.6	15.0	−0.4	−1.4	25.8 – 72.2	24
Station Group 6	67.0	3.3	−0.3	−0.5	60.3 – 73.3	24
Transect I	44.2	17.0	0.2	−1.2	16.4 – 74.8	36
Transect II	51.4	12.1	0.1	−1.3	32.3 – 72.2	36
Transect III	51.5	20.5	−1.5	0.8	2.9 – 70.3	36
Tranect IV	37.1	19.3	0.6	−0.9	9.8 – 71.6	42

SEDIMENT CHEMISTRY (1977)

	Mean	SD	Skewness	Kurtosis	Range	N
Sediment total organic carbon (% organic carbon/dry weight sediment)	0.85	0.32	−0.43	0.08	0.10 – 1.32	75
Station Group 1	0.48	0.34	0.12	−1.57	0.08 – 1.04	12
Station Group 2	0.62	0.18	−0.72	−0.19	0.30 – 0.85	9
Station Group 3	0.88	0.18	−0.51	−0.30	0.50 – 1.19	18
Station Group 4	0.94	0.22	−0.38	−0.67	0.54 – 1.25	12
Station Group 5	0.93	0.25	0.14	−0.98	0.56 – 1.32	12
Station Group 6	1.17	0.26	−1.05	2.75	0.55 – 1.62	12
Sediment Delta ^{13}C (per mil deviations from the PDB standard)	−19.95	0.45	0.82	0.49	(−20.60) – (−18.90)	75
Station Group 1	−19.34	0.46	0.28	−0.21	(−19.93) – (−18.44)	12
Station Group 2	−19.69	0.37	−0.74	0.10	(−20.40) – (−19.19)	9
Station Group 3	−19.99	0.31	0.26	−1.43	(−20.42) – (−19.50)	18
Station Group 4	−20.19	0.26	0.17	−0.82	(−20.58) – (−19.73)	12
Station Group 5	−20.11	0.36	0.53	−1.09	(−20.55) – (−19.52)	12
Station Group 6	−20.29	0.22	−0.43	−0.27	(−20.70) – (−19.93)	12

Table A-5—*Continued*

Variable	Mean	SD	Skew.	Kurt.	95% Empir. CI		N
Winter	−20.12	0.33	1.02	1.18	(−20.60)−	(−19.20)	25
Spring	−20.01	0.46	0.87	−0.01	(−20.58)−	(−18.90)	25
Fall	−19.73	0.48	0.48	1.02	(−20.70)−	(−18.44)	25
Transect I	−19.99	0.46	0.48	−1.13	(−20.58)−	(−19.15)	18
Transect II	−20.16	0.41	0.95	0.20	(−20.70)−	(−19.19)	18
Transect III	−19.82	0.45	1.86	4.37	(−20.40)−	(−18.44)	18
Transect IV	−19.85	0.44	0.65	−0.58	(−20.50)−	(−18.90)	21
HIGH MOLECULAR-WEIGHT HYDROCARBONS (sediment; 1975–1977)							
Total hydrocarbons (μg/g)							
($n-C_{14}$ to $n-C_{32}$)	0.49	0.38	1.72	4.37	0.04−	1.45	85
Pristane:phytane	2.55	2.08	2.34	6.87	0.15−	8.62	68
Pristane:$n-C_{17}$	0.55	0.32	2.06	5.41	0.18−	1.50	73
Phytane:$n-C_{18}$	0.31	0.21	1.48	3.32	0.04−	0.77	68
(Pristane + phytane):n-alkanes	0.02	0.01	1.02	0.86	0.00−	0.06	73
SUM LOW (relative percent)							
(nC_{14} to $n-C_{18}$)	9.78	8.34	1.77	6.54	0.00−	29.62	119
Winter	4.76	5.26	1.28	1.58	0.00−	21.15	35
Spring	11.86	7.93	0.78	1.01	0.00−	35.85	35
Fall	13.22	9.30	2.64	10.63	0.00−	54.86	37
SUM MID (relative percent)							
($n-C_{19}$ to $n-C_{24}$)	23.79	12.97	1.05	0.67	6.72−	57.10	119
1975	35.39	12.90	0.52	−0.43	15.69−	66.92	34
1976	20.32	9.76	1.40	2.44	7.55−	47.14	73
SUM HIGH (relative percent)							
($n-C_{25}$ to $n-C_{32}$)	66.43	14.96	−0.75	0.65	20.15−	90.82	119
1975	55.36	15.67	−0.10	−0.42	19.91−	83.30	34
1976	69.32	11.66	−1.13	3.76	20.15−	90.82	73

Benthic living characteristics

MICROBIOLOGY (1977)

	Mean	SD	Skewness	Kurtosis	Range	N
Fungal counts (CFU/ml)	237.50	408.10	2.68	7.49	34.56–440.44	18
Fungal oil degraders (CFU/ml)	157.22	249.39	2.01	4.33	33.20–281.24	18
Total bacteria (no./ml)	477,986.11	322,844.14				36
Station 1	788,083.33	257,882.45	0.51	−0.26	402,000.0 – 1,310,000.0	12
Station 2	430,083.33	194,982.73	−0.06	0.43	115,000.0 – 758,400.0	12
Station 3	215,791.67	211,454.58	0.32	−0.54	46,300.0 – 632,000.0	12
Bacteria oil degraders (no./ml)	9,725.83	21,808.65	1.17	0.14		36
Winter	2,665.83	2,971.58	3.84	15.17		12
Spring	3,070.00	3,830.13	1.15	0.33	80.0 – 9,100.0	12
Fall	23,441.67	34,379.76	1.92	3.64	130.0 – 13,000.0	12
			2.05	3.41	1,300.0 – 110,000.0	

MEIOFAUNA (1976–1977)

	Mean	SD	Skewness	Kurtosis	Range	N
Total meiofauna species (species/10 cm^2)	4.65	1.67	0.74	0.53	2.0 – 8.5	186
Station Group 1	6.7	1.7	0.2	−0.03	3.5 – 10.5	24
Station Group 2	5.5	1.8	0.6	0.21	2.5 – 9.5	18
Station Group 3	4.4	1.1	1.1	1.61	2.5 – 7.0	36
Station Group 4	4.1	1.2	0.8	1.26	2.0 – 7.5	24
Station Group 5	4.2	1.4	0.4	−0.56	2.0 – 7.3	24
Station Group 6	4.0	1.3	0.4	−0.42	1.5 – 6.8	24
Total meiofauna density (individuals/10 cm^2)	202.86	311.04	2.50	6.09	8.8 – 1,153.3	186
Station Group 1	710.3	408.7	0.2	−1.1	77.8 – 1,447.0	24
Station Group 2	347.5	432.7	1.9	4.3	19.5 – 1,682.5	18
Station Group 3	140.5	193.6	4.0	19.2	23.5 – 1,118.0	36
Station Group 4	91.0	81.4	1.8	3.2	11.5 – 339.3	24
Station Group 5	60.7	38.9	0.8	−0.4	8.5 – 139.5	24
Station Group 6	42.0	26.7	1.0	0.7	8.8 – 112.7	24
Transect I	211.1	264.5	1.9	3.3	8.5 – 1,118.0	36
Transect II	69.0	10.2	2.1	4.4	12.5 – 283.8	36
Transect III	248.8	412.6	2.1	3.1	8.8 – 1,447.0	36
Transect IV	336.9	402.6	1.6	2.0	14.3 – 1,213.0	42

Table A-5—Continued

Variable	Mean	SD	Skew.	Kurt.	95% Empir. CI		N
Nematode density (individuals/10 cm²)							
Station Group 1	152.17	246.83	2.60	6.70	4.5 –	903.5	186
Station Group 2	578.6	338.1	0.1	−0.8	26.5 –	1,267.5	24
Station Group 3	268.5	348.4	1.8	3.6	7.8 –	1,313.5	18
Station Group 4	91.4	109.8	3.1	11.3	8.0 –	580.5	36
Station Group 5	66.9	55.8	1.6	2.7	8.3 –	238.0	24
Station Group 6	46.4	32.2	0.8	−0.5	3.3 –	109.5	24
Transect I	26.7	19.6	1.1	0.4	4.5 –	75.5	24
Transect II	158.6	200.5	1.5	1.2	3.3 –	731.0	36
Transect III	45.5	41.4	2.7	9.7	4.5 –	226.0	36
Transect IV	189.1	329.7	2.2	3.8	6.5 –	1,267.5	36
	267.1	331.1	1.5	1.5	8.5 –	992.5	42
Harpacticoid density (individuals/10 cm²)							
Station Group 1	6.35	15.26	4.92	29.75	0.0 –	56.0	186
Station Group 2	24.5	30.3	2.1	5.5	0.0 –	130.0	24
Station Group 3	8.3	11.4	3.0	10.6	0.0 –	49.5	18
Station Group 4	2.7	4.8	3.5	12.5	0.0 –	23.3	36
Station Group 5	2.5	5.1	3.9	16.1	0.0 –	24.5	24
Station Group 6	1.9	2.5	2.9	10.7	0.0 –	11.8	24
Transect I	1.5	1.8	1.4	0.8	0.0 –	6.0	24
Transect II	4.2	7.3	3.2	11.2	0.00 –	36.3	36
Transect III	1.9	1.8	1.3	1.6	0.00 –	7.5	36
Transect IV	11.8	25.7	3.4	12.9	0.00 –	130.0	36
	7.9	13.7	2.9	9.3	0.00 –	49.5	42
MACROINVERTEBRATES (1976–1977)							
Infauna species (species/0.1 m²)	70.63	41.82	1.72	2.63	26.0 –	201.0	186
Station Group 1	129.9	61.6	−0.01	−1.51	38.0 –	226.0	24
Station Group 2	73.5	46.9	0.98	−0.81	32.0 –	161.0	18
Station Group 3	56.6	20.2	0.81	0.53	26.0 –	115.0	36

Overall Base Line Results [195]

Station Group 4	66.6		1.21	20.0	169.0	24
Station Group 5	76.5	39.5	-0.87	29.0	133.0	24
Station Group 6	57.1	31.7	1.11	14.0	91.0	24
Transect I	73.4	17.4	-0.27	34.0	156.0	36
Transect II	51.2	26.3	1.58	27.0	107.0	36
Transect III	67.6	18.7	1.24	20.0	207.0	36
Transect IV	105.1	50.3	2.62	14.0	206.0	42
		52.5	-0.56			
Infauna density						
(individuals/0.1 m^2)	667.53	1,077.35	3.36	47.0	4,475.0	186
Station Group 1	2,726.3	1,834.9	0.64	491.0	6,770.0	24
Station Group 2	922.3	582.1	1.08	323.0	2,249.0	18
Station Group 3	404.9	332.8	2.75	62.0	1,877.0	36
Station Group 4	255.0	259.8	2.37	43.0	1,063.0	24
Station Group 5	290.8	160.9	0.48	58.0	639.0	24
Station Group 6	167.3	71.2	0.18	37.0	324.0	24
Transect I	801.5	959.2	2.14	69.0	4,303.0	36
Transect II	281.4	181.3	1.22	37.0	884.0	36
Transect III	1,043.4	1,886.6	2.17	50.0	6,772.0	36
Transect IV	885.2	952.5	1.49	37.0	3,084.0	42
Epifauna species (species/trawl)	8.58	4.43	0.99	1.0	19.0	185
Epifauna density (individuals/trawl)	136.34	218.06	3.78	2.0	823.0	185
DEMERSAL FISHES (1976–1977)						
Fish species (species/trawl)	14.30	5.64	0.52	4.0	27.0	185
Fish density (individuals/trawl)	126.69	167.82	4.20	8.0	566.0	185
Fish biomass (g/trawl)	2,708.76	2,630.03	3.18	175.1	8,142.7	185

Table A-5—*Continued*

Variable	Mean	SD	Skew.	Kurt.	95% Empir. CI	N
HIGH MOLECULAR-WEIGHT HYDROCARBON BODY BURDENS (1975–1977)						
Penaeus aztecus						
Total hydrocarbons (µg/g) (n–C_{14} to n–C_{32})	0.15	0.28	4.70	25.95	0.00– 1.77	46
Lutjanus campechanus–gonad						
Total hydrocarbons (µg/g) (n–C_{14} to n–C_{32})	36.79	31.76	0.95	0.20	1.74– 98.56	9
Lutjanus campechanus–gill						
Total hydrocarbons (µg/g) (n–C_{14} to n–C_{32})	4.67	6.06	1.87	3.72	0.00– 20.01	11
Lutjanus campechanus–liver						
Total hydrocarbons (µg/g) (n–C_{14} to n–C_{32})	8.68	9.55	2.90	9.88	1.13– 43.80	20
SUM MID (relative percent) (n–C_{19} to n–C_{24})	6.58	7.77	1.09	0.12	0.00– 24.10	20
1976	0.90	1.48	2.10	4.79	0.00– 4.80	11
1977	13.51	6.51	0.59	-1.09	6.00– 24.10	9
Average OEP (odd-even preference) (n–C_{14} to n–C_{32})	6.59	6.35	0.67	-1.70	1.13– 15.83	14
1976	14.70	1.25	-1.04	-0.35	12.81– 15.83	5
1977	2.09	0.87	0.69	-1.62	1.13– 3.31	9
Lutjanus campechanus–muscle						
Total hydrocarbons (µg/g) (n–C_{14} to n–C_{32})	0.68	1.58	4.19	18.43	0.00– 7.40	21
Rhomboplites aurorubens–gill						
Total hydrocarbons (µg/g)						

Overall Base Line Results [197]

Rhomboplites aurorubens—gonad							
Total hydrocarbons (μg/g) (n-C_{14} to n-C_{32})	6.85	7.24	2.56	6.89	2.32–	25.31	9
Rhomboplites aurorubens—muscle							
Total hydrocarbons (μg/g) (n-C_{14} to n-C_{32})	1.37	1.30	1.17	0.59	0.02–	4.38	20
Rhomboplites aurorubens—liver							
Total hydrocarbons (μg/g) (n-C_{14} to n-C_{32})	13.60	9.78	0.84	0.08	0.57–	35.85	18
Pristane: n-C_{17}	10.67	9.92	1.65	2.02	1.85–	36.00	15
Winter	7.03	0.00	0.00	0.00	7.03–	7.03	1
March	6.31	0.00	0.00	0.00	6.31–	6.31	1
April	36.00	0.00	0.00	0.00	36.00–	36.00	1
Spring	20.60	7.22	0.00	0.00	12.55–	26.08	3
July	5.78	1.41	–1.42	0.00	4.27–	7.07	3
August	5.29	2.33	–0.68	0.00	3.64–	6.94	2
Fall	8.46	2.65	0.00	0.00	6.59–	10.34	2
November	2.18	0.00	0.00	0.00	2.18–	2.18	1
December	1.85	0.00	0.00	0.00	1.85–	1.85	1
Trachurus lathami							
Total hydrocarbons (μg/g) (n-C_{14} to n-C_{32})	8.58	12.00	2.47	6.60	0.03–	51.32	25
Stenotomus caprinus							
Total hydrocarbons (μg/g) (n-C_{14} to n-C_{32})	1.05	1.85	3.32	11.70	0.02–	8.58	26
Loligo pealei							
Total hydrocarbons (μg/g) (n-C_{14} to n-C_{32})	1.93	3.35	2.58	6.30	0.00–	14.09	44
Serranus atrobranchus							

[198] Appendix A

Table A-5—*Continued*

Variable	Mean	SD	Skew.	Kurt.	95% Empir. CI	N
Total hydrocarbons (µg/g) (n–C_{14} to n–C_{32}) *Pristipomoides aquilonaris*	0.19	0.21	0.88	−0.47	0.00– 0.68	26
Total hydrocarbons (µg/g) (n–C_{14} to n–C_{32})	3.04	4.33	2.01	3.27	0.12– 15.80	37
TRACE METAL BODY BURDENS (1975–1977)						
Penaeus aztecus–flesh						
Zinc (ppm dry weight)	52.90	8.79	−0.83	2.73	20.00– 68.00	51
Cadmium (ppm dry weight)	0.11	0.08	0.83	0.36	0.01– 0.25	51
Trachurus lathami–flesh						
Cadmium (ppm dry weight)	0.09	0.09	1.20	0.33	0.01– 0.30	24
Calcium (ppm dry weight)	882.5	498.97	2.37	7.61	310.00– 2,500.00	16
Aluminum (ppm dry weight)	22.75	11.20	1.11	0.84	10.00– 50.00	16
Serranus atrobranchus–flesh						
Zinc (ppm dry weight)	10.88	3.28	−0.72	1.46	2.00– 17.00	24
Stenotomus caprinus–flesh						
Cadmium (ppm dry weight)	0.08	0.05	0.45	−0.82	0.02– 0.16	17
Lutjanus campechanus–gill						
Vanadium (ppm dry weight)	0.39	0.26	0.55	−3.10	0.15– 0.70	5

[a] The analytic methods used to determine the values for the variables presented herein are detailed in Flint and Rabalais (1980).

APPENDIX B
MAPS OF VARIABLES'
GEOGRAPHIC DISTRIBUTIONS

The purpose of this appendix is to present a quick reference to the distributional characteristics of those environmental variables measured during the south Texas outer continental shelf (STOCS) study that showed significant ($P < 0.05$) spatial variation over the study area. Presented are the mean and 95% normal confidence interval statistics for every station on the Texas shelf sampled over the study period (1975–1977). The intention of this presentation is to provide decision makers and environmental managers with a quick reference to the study area by the variables included so that decisions concerning the management of the ecosystem can be made or criteria for further monitoring of the system can be developed.

Sampling Scheme

The variables presented represent several different sampling schemes. For some variables, data were collected in all three years of the study (1975–1977); for others, only in one or two years of study. For further reference concerning the extent of sampling, the reader should see the specific scientific section concerning the certain variable in Flint and Rabalais (1980). The means and confidence intervals presented in the distributional maps of this appendix represent data collected during the three meteorological seasons of each year only: winter, spring, and fall.

Spatially, two different sampling schemes are presented in the maps: (1) a 12-station scheme involving Stations 1–3, Transects I–IV, primarily for pelagic sampling, and (2) a 25-station scheme involving

[200] Appendix B

Stations 1–6, Transects I–III, and Stations 1–7, Transect IV, primarily for benthic sampling (see Chapter 1, Figure 3). Table 1 (Chapter 1) lists the LORAN and LORAC coordinates, as well as the latitude, longitude, and water depth of each site represented by one of the two sampling schemes described above.

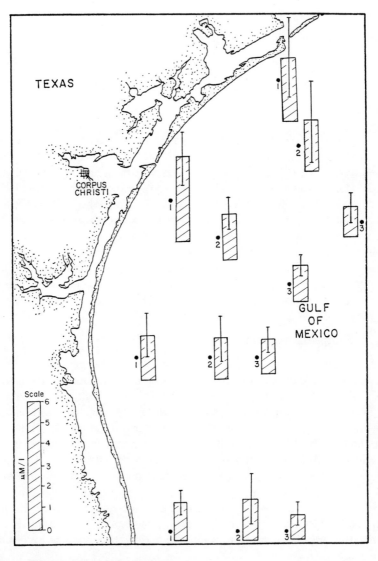

Figure B-1. Surface water silicate.

Variables' Geographic Distributions [201]

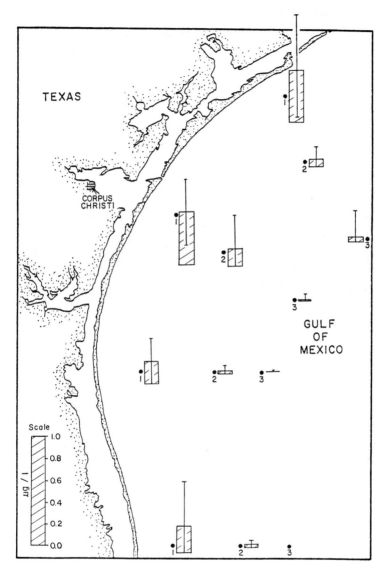

Figure B-2. Surface net chlorophyll.

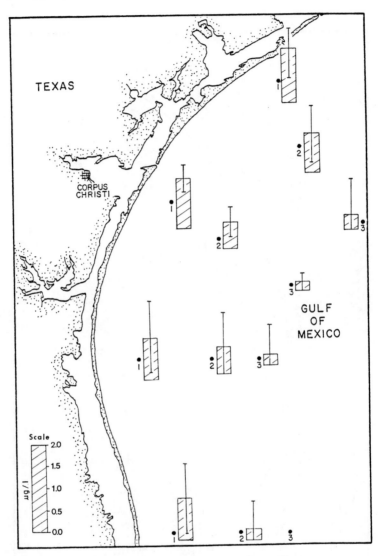

Figure B-3. Surface net phaeophytin.

Variables' Geographic Distributions [203]

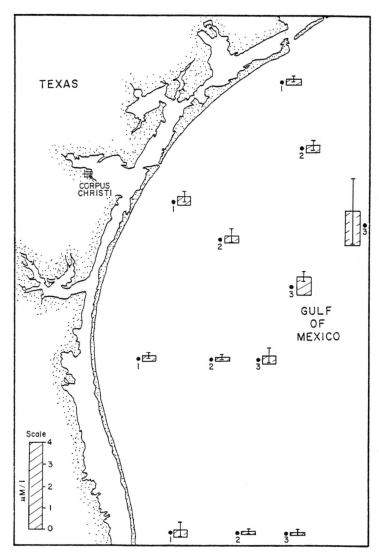

Figure B-4. Bottom water phosphate.

[204] Appendix B

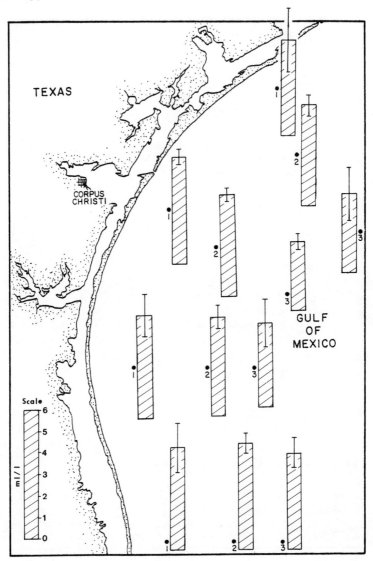

Figure B-5. Bottom water dissolved oxygen.

Variables' Geographic Distributions [205]

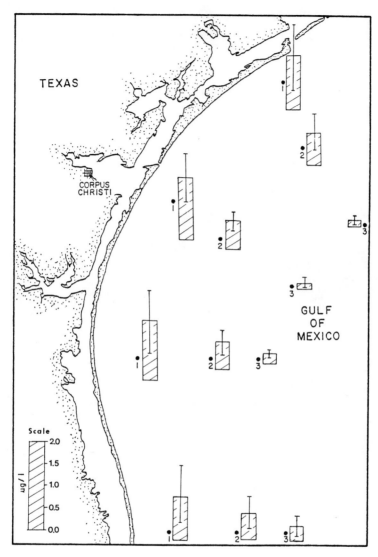

Figure B-6. Bottom water total chlorophyll.

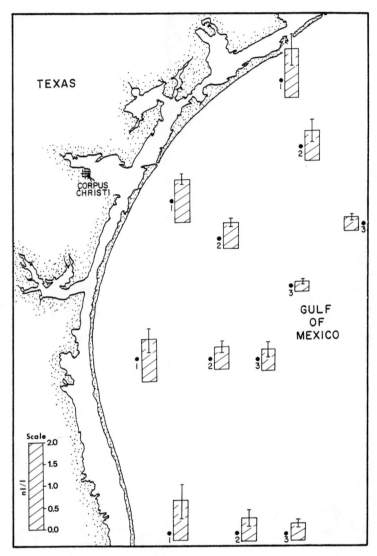

Figure B-7. Bottom water propene.

Variables' Geographic Distributions [207]

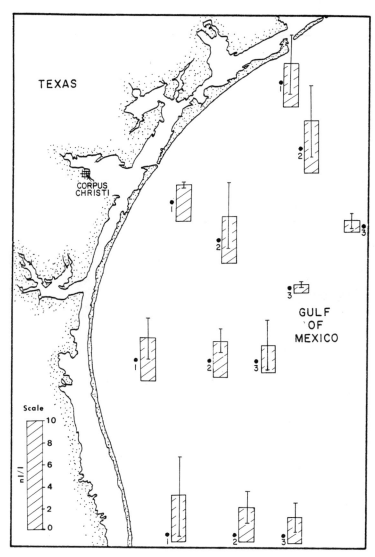

Figure B-8. Bottom water ethene.

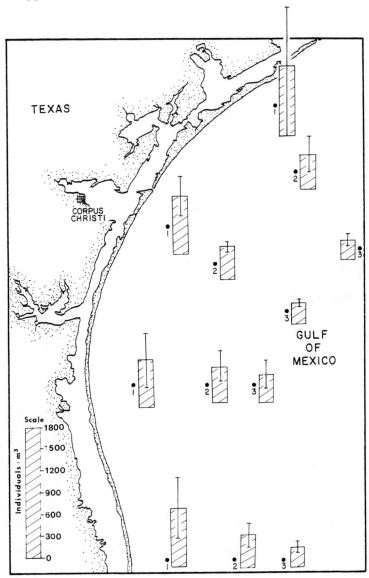

Figure B-9. Copepod total density.

Variables' Geographic Distributions [209]

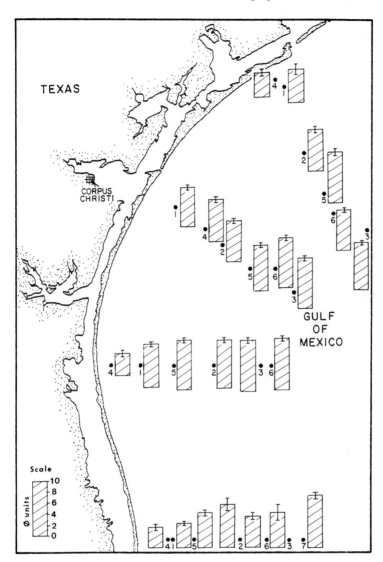

Figure B-10. Sediment mean grain size.

[210] Appendix B

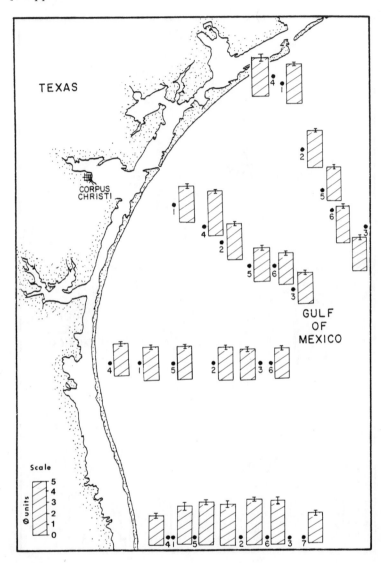

Figure B-11. Sediment grain size standard deviation.

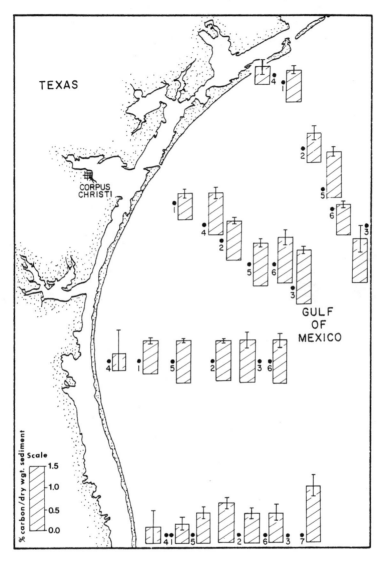

Figure B-12. Sediment total organic carbon.

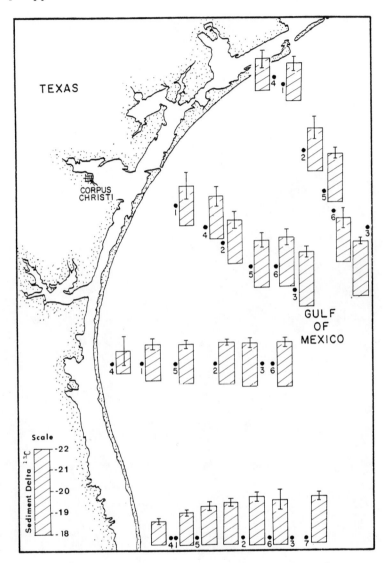

Figure B-13. Sediment Delta ^{13}C.

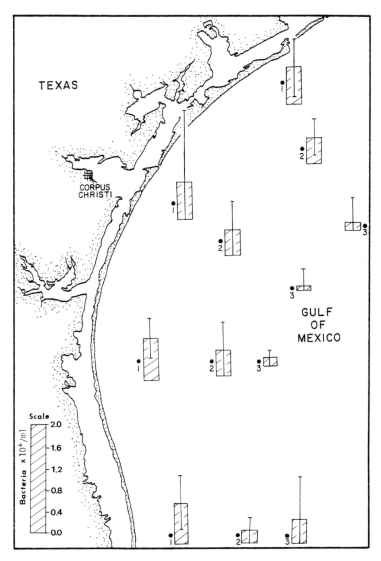

Figure B-14. Total sediment bacteria.

[214] Appendix B

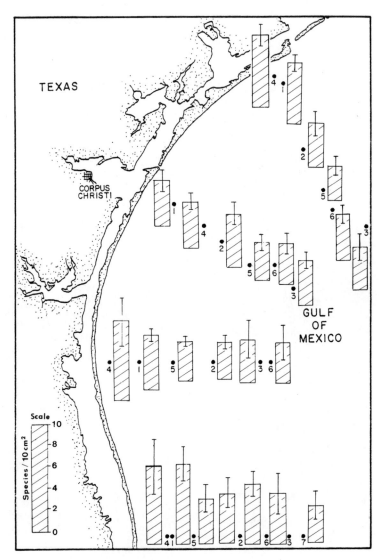

Figure B-15. Total meiofauna species.

Variables' Geographic Distributions [215]

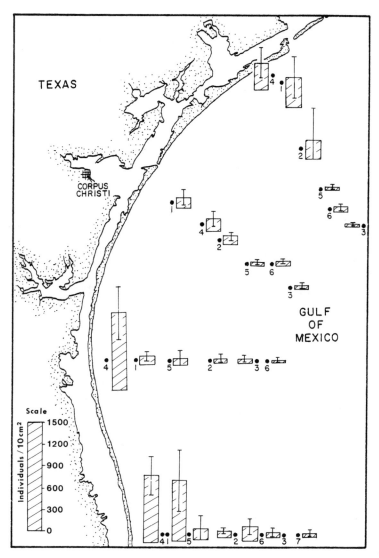

Figure B-16. Total meiofauna density.

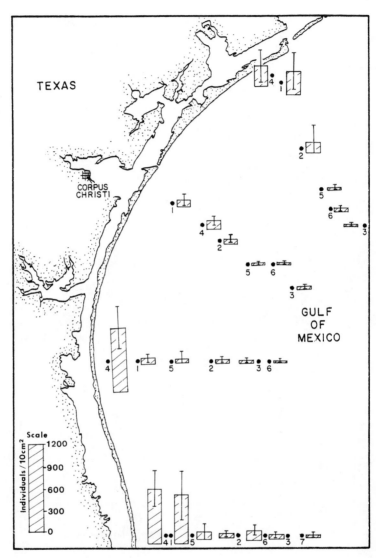

Figure B-17. Nematode density.

Variables' Geographic Distributions [217]

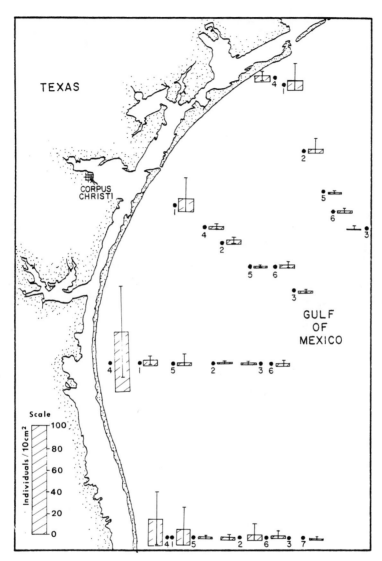

Figure B-18. Harpacticoid density.

[218] Appendix B

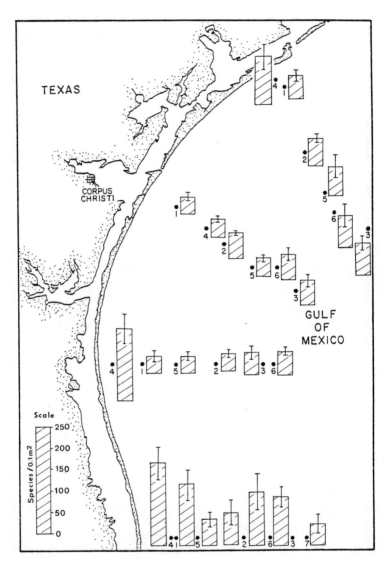

Figure B-19. Infauna species.

Variables' Geographic Distributions [219]

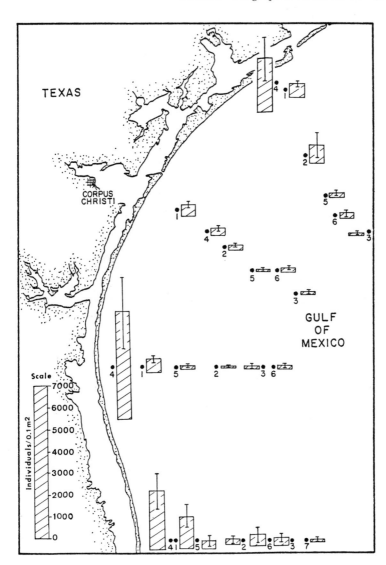

Figure B-20. Infauna density.

REFERENCES

Armstrong, R. 1976. Historical temperature and salinity data. In *Environmental studies of the south Texas outer continental shelf, 1975,* ed. J. W. Angelovic, pp. 20–21. Vol. II. National Oceanic and Atmospheric Administration final report to the Bureau of Land Management, Department of the Interior, Washington, D.C. Interagency agreement #08550-IA5-19.

Bell, S. S., and Coull, B. C. 1978. Field evidence that shrimp predation regulates meiofauna. *Oecologia (Berl.)* 35:141–148.

Berryhill, H. L., Jr. 1977. *Environmental studies, south Texas outer continental shelf, 1975: An atlas and integrated summary.* Final report to the Bureau of Land Management, Department of the Interior, Washington, D.C.

Berryhill, H. L., Jr., ed. 1978. *Environmental studies, south Texas outer continental shelf, geology, 1977.* Final report to the Bureau of Land Management, Department of the Interior, Washington, D.C. Contract AA550-MU7-27.

Blumer, M., Mullin, M. M., and Thomas, D. W. 1964. Pristane in the marine environment. *Helgolander Wiss. Meeresunters.* 10:187–201.

Blumer, M., and Snyder, W. D. 1965. Isoprenoid hydrocarbons in recent sediments: Presence of pristane and probable absence of phytane. *Science* 150:1588–1589.

Boesch, D. F. 1973. Classification and community structure of macrobenthos in the Hampton Roads Area, Virginia. *Mar. Biol.* 21:226–244.

———. 1979. *Benthic ecological studies: Macrobenthos.* Special Report in Applied Marine Science and Ocean Engineering #194. Gloucester Point: Virginia Institute of Marine Science.

Bovée, F. de, and Soyer, J. 1977. Le meiobenthos des Iles Kerguelan. Données quantitatives. II. Le plateau continental. *Comité National Français des Recherches Antarctiques* 42:249–258.

Bowman, T. E. 1971. The distribution of calanoid copepods off the southeastern United States between Cape Hatteras and southern Florida. *Smithsonian Contributions to Zoology* 96:1–58.

References

Breuer, J. P. 1962. An ecological survey of the lower Laguna Madre of Texas, 1953-1959. *Publications of the Institute of Marine Science, University of Texas* 8:153-183.

Brooks, R. R. 1977. Pollution through trace elements. In *Environmental chemistry*, ed. J. O. Bockris, pp. 429-476. New York: Plenum Press.

Buzas, M. A. 1978. Foraminifera as prey for benthic deposit feeders: Results of predator exclusion experiments. *J. Mar. Res.* 36:617-625.

Calder, J. A. 1977. Seasonal variation of hydrocarbons in the water column of the MAFLA lease area. In *Fate and effects of petroleum hydrocarbons in marine organisms and ecosystems*, ed. D. A. Wolfe, pp. 432-441. New York: Pergamon Press.

Carpenter, E. J., and McCarthy, J. J. 1978. Benthic nutrient regeneration and high rate of primary production in continental shelf waters. *Nature* 274:188-189.

Cerniglia, C. E., Hebert, R. L., Szaniszlo, P. J., and Gibson, D. T. 1978. Fungal transformation of naphthalene. *Arch. Microbiol.* 117:135-143.

Chapman, J. 1979. *Statistical report of the pollen and mold committee, 1978.* Columbus, Ohio: American Academy of Allergy.

Chittenden, M. E., and McEachran, J. D. 1976. Composition, ecology, and dynamics of demersal fish communities on the northwestern Gulf of Mexico continental shelf, with a similar synopsis for the entire Gulf. Technical report for the Center for Marine Resources, Texas A&M University, College Station, Texas.

Chittenden, M. E., and Moore, D. 1977. Composition of the ichthyofauna inhabiting the 100-meter bathymetric contour of the Gulf of Mexico, Mississippi River to the Rio Grande. *Northeast Gulf Science* 1:106-114.

Chua, K. E., and Brinkhurst, R. O. 1973. Bacteria as potential nutritional resources for three sympatric species of tubificid oligochaetes. In *Estuarine microbial ecology*, ed. L. H. Stevenson and R. R. Colwell, pp. 512-517. Columbia: Univ. of South Carolina Press.

Clark, R. C. 1974. Methods for establishing levels of petroleum contamination in organisms and sediment as related to marine pollution monitoring. In *Marine Pollution Monitoring (Petroleum)*, pp. 189-194. National Bureau of Standards Special Publication 409.

Clark, R. C., and Blumer, M. 1967. Distribution of n-paraffins in marine organisms and sediment. *Limnol. Oceanogr.* 12:79-87.

Clark, R. C., and Finley, F. S. 1973. Techniques for analysis of paraffin hydrocarbons and for interpretation of data to assess oil spill effects in aquatic organisms. In *Proceedings of the 1973 Joint Conference of Prevention and Control of Oil Spills*, pp. 161-172. Washington, D.C., American Petroleum Institute.

Cohen, J. 1969. *Statistical power analysis for the behavioral sciences.* New York: Academic Press.

Conover, R. J. 1971. Some relations between zooplankton and bunker C oil in Chedabucto Bay following the wreck of the tanker *Arrow. J. Fish. Res. Board Can.* 28:1327-1330.

Cooper, J. E., and Bray, E. E. 1963. A postulated role of fatty acids in petroleum formation. *Geochim. Cosmochim. Acta* 27:1113–1127.
Copeland, B. J. 1965. Fauna of Aransas Pass Inlet, Texas. I. Emigration as shown by tide trap collections. *Publications of the Institute of Marine Science, University of Texas* 10:9–12.
Coull, B. C. 1973. Estuarine meiofauna: A review. Trophic relationships and microbial interactions. In *Estuarine microbial ecology*, ed. L. H. Stevenson and R. R. Colwell, pp. 499–511. Columbia: Univ. South Carolina Press.
Cuzon du Rest, R. P. 1963. Distribution of zooplankton in the salt marshes of southeastern Louisiana. *Publications of the Institute of Marine Science, University of Texas* 9:132–155.
Dawson, C. E. 1964. A revision of the western Atlantic flatfish genus *Gymnachirus* (the naked soles). *Copeia* 1964:646–665.
Day, J. S., Field, J. G., and Montgomery, M. 1971. Use of numerical methods to determine the distribution of benthic fauna across the continental shelf of North Carolina. *J. Anim. Ecol.* 40:93–126.
Devine, M. 1976. Tides, tidal currents and sea level. In *Environmental studies of the south Texas outer continental shelf, 1975*, ed. J. W. Angelovic, pp. 23–26. Vol. II. National Oceanic and Atmospheric Administration final report to the Bureau of Land Management, Department of the Interior, Washington, D.C. Interagency agreement #08550-IA5-19.
Dickie, T. M. 1972. Food chains and fish production. *International Commission for the Northwest Atlantic Fisheries Special Publication* 8:201–221.
Dulka, J. J., and Risby, T. H. 1976. Ultratrace metals in some environmental and biological systems. *Anal. Chem.* 48:640a–653a.
Ehrhardt, M., and Blumer, M. 1972. The source identification of marine hydrocarbons by gas chromatography. *Environ. Pollut.* 3:179–194.
Farrington, J. W., Giam, C. S., Harvey, G. R., Parker, P. L., and Teal, J. 1972. Analytical techniques for selected organic compounds. In *Marine pollution monitoring: Strategies for a national program*, pp. 152–176. Santa Catalina, California: Santa Catalina Marine Biology Laboratory.
Fauchald, K., and Jumars, P. A. 1979. The diet of worms: A study of polychaete feeding guilds. *Oceanography and Marine Biology Annual Review* 17:193–284.
Field, J. G. 1971. A numerical analysis of changes in the soft-bottom fauna along a transect across False Bay, South Africa. *J. Exp. Mar. Biol. Ecol.* 7:215–253.
Finucane, J. 1976. Ichthyoplankton. In *Environmental studies of the south Texas outer continental shelf, 1975*, pp. 20–31. Vol. I. National Oceanic and Atmospheric Administration final report to the Bureau of Land Management, Department of the Interior, Washington, D.C.
———. 1977. Ichthyoplankton. In *Environmental studies of the south Texas outer continental shelf, 1976*, pp. 2–238, 296–485. National Oceanic and Atmospheric Administration final report to the Bureau of Land Management, Department of the Interior, Washington, D.C.
Flint, R. W., and Griffin, C. W., eds. 1979. *Environmental studies, south Texas*

outer continental shelf, biology and chemistry. The 1977 final report to the Bureau of Land Management, Department of the Interior, Washington, D.C. Contract AA550-CT7-11.

Flint, R. W., and Rabalais, N. N., eds. 1980. *Environmental studies, south Texas outer continental shelf, 1975–1977.* Vol. III. Final report to the Bureau of Land Management, Department of the Interior, Washington, D.C. Contract AA551-CT8-51.

Fry, B. D. 1977. Stable carbon isotope ratios—A tool for tracing food chains. M.S. thesis, University of Texas, Austin.

Galtsoff, P. S., ed. 1954. Gulf of Mexico: Its origin, waters, and marine life. *Fish. Bull.* 55(89):1–604.

Gerlach, S. A. 1971. On the importance of marine meiofauna for benthos communities. *Oecologia (Berl.)* 6:176–190.

⸻. 1978. Food-chain relationships in subtidal silty sand marine sediments and the role of meiofauna in stimulating bacterial productivity. *Oecologia (Berl.)* 33:55–69.

Gilfillan, E. S., Mayo, D. W., Page, D. S., Donovan, D., and Hanson, S. 1977. Effects of varying concentrations of petroleum hydrocarbons in sediments on carbon flux in *Mya arenaria.* In *Physiological responses of marine biota to pollutants,* ed. F. J. Vernberg et al., pp. 299–314. New York: Academic Press.

Glemarec, J. 1973. The benthic communities of the European North Atlantic continental shelf. *Oceanography and Marine Biology Annual Review* 11:263–289.

Grassle, J. F., and Grassle, J. P. 1974. Opportunistic life histories and genetic systems in marine benthic polychaetes. *J. Mar. Res.* 32:253–284.

Griffin, C. W., ed. 1979. *Environmental studies, south Texas outer continental shelf, biology and chemistry.* The supplemental report to the 1976 final report to the Bureau of Land Management, Department of the Interior, Washington, D.C. Contract AA550-CT6-17.

Groover, R. D., ed. 1977a. *Environmental studies, south Texas outer continental shelf, biology and chemistry.* The 1976 final report to the Bureau of Land Management, Department of the Interior, Washington, D.C. Contract AA550-CT6-17.

Groover, R. D., ed. 1977b. *Environmental studies, south Texas outer continental shelf, rig monitoring.* The 1976 final report to the Bureau of Land Management, Department of the Interior, Washington, D.C. Contract AA550-CT6-17.

Gunter, G. 1945. Marine fishes of Texas. *Publications of the Institute of Marine Science, University of Texas* 1:1–190.

Hann, R. W., Jr., and Slowey, J. F. 1972. Sediment analysis Galveston Bay. Technical report 24, Environmental Engineering Division, Texas A&M University, College Station, Texas.

Hedgpeth, J. W. 1953. An introduction to the zoogeography of the northwestern Gulf of Mexico with reference to the invertebrate fauna. *Publications of the Institute of Marine Science, University of Texas* 3:107–224.

Hildebrand, H. H. 1954. A study of the fauna of the brown shrimp (*Penaeus aztecus* Ives) grounds in the western Gulf of Mexico. *Publications of the Institute of Marine Science, University of Texas* 3:229–366.

Holmes, C. W., Slade, E. A., and McLerran, C. J. 1974. Migration and redistribution of zinc and cadmium in marine estuarine systems. *Environ. Sci. Technol.* 8:255–259.

International Council for Exploration of the Sea. 1978. *On the feasibility of effects monitoring.* ICES Cooperative Research Report No. 75.

Jacobs, W. C. 1951. The energy exchange between sea and atmosphere and some of its consequences. *Bulletin of Scripps Institute of Oceanography, University of California* 6(2):27–122.

Johnson, T. W., Jr., and Sparrow, F. K., Jr. 1961. *Fungi in oceans and estuaries.* New York: J. Cramer.

Jones, E. B. G. 1976. *Recent advances in aquatic mycology.* New York: John Wiley and Sons.

Jones, J. S. 1950. Bottom fauna communities. *Biological Review* 25:283–313.

Jones, R. S., Copeland, B. J., and Hoese, H. D. 1965. A study of the hydrography of inshore waters in the western Gulf of Mexico off Port Aransas, Texas. *Publications of the Institute of Marine Science, University of Texas* 10:22–32.

Kamykowski, D., and Batterton, J. 1979. Biological characterization of the nepheloid layer. In *Environmental studies, south Texas outer continental shelf, biology and chemistry,* ed. C. W. Griffin, 11-1–11-57. The supplemental report to the 1976 final report to the Bureau of Land Management, Department of the Interior, Washington, D.C. Contract AA550-CT6-17.

Kamykowski, D., Pulich, W. M., and Van Baalen, C. 1977. Phytoplankton and productivity. In *Environmental studies, south Texas outer continental shelf, biology and chemistry,* ed. R. D. Groover, 4-9–4-39. The 1976 final report to the Bureau of Land Management, Department of the Interior, Washington, D.C. Contract AA550-CT6-17.

Kerlinger, F. N., and Pedhazur, E. J. 1973. *Multiple regression in behavioral research.* New York: Holt, Rinehart, and Winston.

Koons, C. B. 1977. Distribution of volatile hydrocarbons in some Pacific ocean waters. *Proceedings 1977 Oil Spill Conference, March 8–10, New Orleans, Louisiana,* pp. 589–592. Publication #4284. Washington, D.C.: American Petroleum Institute.

Leipper, D. F. 1954. Physical oceanography of the Gulf of Mexico. In *Gulf of Mexico, its origin, waters, and marine life,* ed. P. S. Galtsoff. *Fish. Bull.* 55(89):119–137.

Levinton, J. 1972. Stability and trophic structure in deposit-feeding and suspension-feeding communities. *Amer. Naturalist* 106:472–486.

Lewis, T. C., and Yerger, R. W. 1976. Biology of five species of sea robins (Pisces, Triglidae) from the northeastern Gulf of Mexico. *Fish. Bull.* 74:93–103.

Lie, U. 1967. A quantitative study of benthic infauna in Puget Sound, Washington, U.S.A. *Fiskeridirektoratets Skrifter, Serie Havundersoekelser* 14:229–556.

McAullife, C. D. 1976. Surveillance of the marine environment for hydrocarbons. *Mar. Sci. Commun.* 2:13–42.

McCall, P. L. 1977. Community patterns and adaptive strategies of the infaunal benthos of Long Island Sound. *J. Mar. Res.* 35:221–266.

McIntyre, A. D. 1964. Meiobenthos of sub-littoral muds. *J. Mar. Biol. Ass. U.K.* 44:665–674.

_____. 1969. Ecology of marine meiobenthos. *Biological Review* 44:245–290.

_____. 1974. Meiobenthos. *Proceedings of the Challenger Society* 4(3):1–9.

Mare, M. 1942. A study of a marine benthic community with special reference to the micro-organisms. *J. Mar. Biol. Ass. U.K.* 25:517–554.

Markovetz, A. J., Jr., Cazin, J., and Allen, J. E. 1968. Assimilation of alkanes and alkenes by fungi. *Applied Microbiology* 16:487–489.

Martin, J. H., and Knauer, G. A. 1973. The elemental composition of plankton. *Geochim. Cosmochim. Acta* 37:1638–1653.

Mattison, G. C. 1948. Bottom configuration in the Gulf of Mexico. *Journal of Coast and Geodetic Survey* 1:76–82.

Menzel, D. W., and Ryther, J. H. 1961. Zooplankton in the Sargasso Sea off Bermuda and its relation to organic production. *Journal du Conseil* 26:260–268.

Middleditch, B. S., and Basile, B. 1978. Determined levels, pathways, and bioaccumulation of selected discharge constituents (non-metals) in the marine ecosystem in the oil field. In *Environmental Assessment of an Active Oil Field in the Northwestern Gulf of Mexico, 1977–1978,* ed. W. B. Jackson, pp. 2.4.1-1–2.4.1-250. Vol. III. National Oceanic and Atmospheric Administration National Marine Fisheries Service Annual Report to the Environmental Protection Agency.

Mills, E. L. 1975. Benthic organisms and the structure of marine ecosystems. *J. Fish. Res. Board Can.* 32:1657–1663.

Mills, E. L., and Fournier, R. O. 1979. Fish production and the marine ecosystems of the Scotian Shelf, eastern Canada. *Mar. Biol.* 54:101–108.

Molander, A. 1928. Animal communities on soft bottom areas in the Gullmar Fjord. *Kristinebergs Zool. Sta. 1877–1927* 2:1–90.

Neff, J. W., Foster, R. S., and Slowey, J. F. 1978. Availability of sediment adsorbed heavy metals to benthos with particular emphasis on deposit-feeding infauna. Technical report D-78-42 to Chief of Engineers, U.S. Army, Washington, D.C. Contract No. DACW-39-57-C-0096.

Newell, R. 1965. The role of detritus in the nutrition of two marine deposit feeders, the prosobranch *Hydrobia ulvae* and the bivalve *Macoma balthica*. *Proceedings of the Zoological Society of London* 144:25–45.

Nichols, F. N. 1978. Infaunal biomass and production of a mudflat, San Francisco Bay, California. In *Benthic ecology,* ed. B. C. Coull, pp. 339–358. Columbia: Univ. of South Carolina Press.

Nie, N. H., Hull, C. H., Jenkins, J. G., Steinbrenner, K., and Bent, D. H. 1975. *Statistical package for the social sciences.* New York: McGraw-Hill.

Nowlin, W. 1971. Water masses and general circulation of the Gulf of Mexico. *Oceanography International* 12:23–33.
Odum, E. P. 1959. *Fundamentals of ecology*. Philadelphia: W. B. Saunders Co.
Orton, R. B. 1964. *The climate of Texas and the adjacent Gulf waters*. Washington, D.C.: U.S. Department of Commerce Weather Bureau.
Parker, P. L., ed. 1976. *Environmental studies, south Texas outer continental shelf, biology and chemistry*. The 1975 final report to the Bureau of Land Management, Department of the Interior, Washington, D.C. Contract 08550-CT5-16.
Parker, R. H. 1960. Ecology and distributional patterns of marine macroinvertebrates, northern Gulf of Mexico. In *Recent sediments, northwest Gulf of Mexico*, ed. F. P. Shepard, F. B. Phleger, and T. H. Van Andel, pp. 302–337. Tulsa, Oklahoma: American Association of Petroleum Geologists.
Parker, R. H., and Curray, J. R. 1956. Fauna and bathymetry of banks of continental shelf, northwest Gulf of Mexico. *Bulletin of the American Association of Petroleum Geologists* 40:2428–2439.
Pequegnat, W. E., and Sikora, W. B. 1977. Meiofauna project. In *Environmental studies, south Texas outer continental shelf, biology and chemistry*, ed. R. D. Groover, pp. 8-1–8-55. The 1976 final report to the Bureau of Land Management, Department of the Interior, Washington, D.C. Contract AA550-CT6-17.
Perkins, E. J. 1958. The food relationships of the microbenthos with particular reference to that found at Whitestable, Kent. *Annals and Magazine of Natural History* 13:64–77.
Peters, D. S., and Kjelson, M. A. 1975. Consumption and utilization of food by various postlarval and juvenile fishes of North Carolina estuaries. In *Estuarine research: Chemistry, biology and the estuarine system*, ed. L. E. Cronin, pp. 448–472. New York: Academic Press.
Petersen, C. G. J. 1913. Valuation of the sea. II. The animal communities of the sea bottom and their importance for marine zoogeography. *Reports of the Danish Biological Station* 25:1–44.
———. 1918. The sea bottom and its production of fish food. *Reports of the Danish Biological Station* 25:1–62.
Pollard, R. T. 1977. Observations and theories of Langmiur circulation and their roles in near surface mixing. *Deep-Sea Res.* Sir George Deacon Univ. Suppl.: 235–251.
Rao, C. R. 1965. *Linear statistical inference and its applications*. New York: Wiley.
Rhoads, D. C. 1974. Organism-sediment relations in the muddy seafloor. *Oceanography and Marine Biology Annual Review* 12:263–300.
Rhoads, D. C., Tenore, K., and Browne, M. 1974. The role of resuspended bottom mud in nutrient cycles of shallow embayments. *Estuarine Res.* 1:563–579.
Rivas, L. 1969. Fisherman's atlas of monthly sea surface temperatures for the Gulf of Mexico. Circular 300. Washington, D.C.: Bureau of Commercial Fishing, U.S. Fish and Wildlife Service.
Rowe, G. T., Clifford, C. H., Smith, K. L., and Hamilton, P. L. 1975. Benthic

nutrient regeneration and its coupling to primary productivity in coastal waters. *Nature* 255:215–217.

Rowe, G. T., Polloni, P. T., and Horner, G. S. 1974. Benthic biomass estimates from the northwestern Atlantic Ocean and northern Gulf of Mexico. *Deep-Sea Res.* 21:641–650.

Rubright, J. S. 1978. An investigation into the role of meiofauna in the food chain of a shrimp mariculture pond system. M.S. thesis, Texas A&M University, College Station, Texas.

Ryther, J. H., and Yentsch, C. S. 1957. The estimation of phytoplankton production in the ocean from chlorophyll and light data. *Limnol. Oceanogr.* 2:281–286.

Saila, S. B. 1976. Sedimentation and food resources: Animal-sediment relations. In *Marine sediment transport and environmental management*, ed. D. J. Stanley and D. J. P. Swift, pp. 479–492. New York: John Wiley and Sons.

Sander, F., and Moore, E. 1978. A comparative study of inshore and offshore copepod populations at Barbados, West Indies. *Crustaceana* 35:225–240.

Sanders, H. L. 1960. Benthic studies in Buzzards Bay. III. The structure of the soft-bottom community. *Limnol. Oceanogr.* 5:138–153.

———. 1968. Marine benthic diversity: A comparative study. *Amer. Naturalist* 102:243–282.

Searle, S. R. 1971. *Linear models*. New York: John Wiley and Sons.

Slowey, J. F., Riddle, D. C., Rising, C. A., and Garrett, R. L. 1973. Natural background levels of heavy metals in Texas estuarine sediments. Report to the Environmental Engineering Division, Texas Water Quality Board.

Snedecor, G. W., and Cochran, W. G. 1967. *Statistical methods*. Ames: Iowa State Univ. Press.

Steele, C. W. 1967. Fungus populations in marine waters and coastal sands of the Hawaiian, Line, and Phoenix Islands. *Pac. Sci.* 21:317–331.

Steele, J. H. 1974. *The structure of marine ecosystems*. Cambridge, Massachusetts: Harvard Univ. Press.

Szaniszlo, P. J. 1979. Water column and benthic microbiology-mycology. In *Environmental studies, south Texas outer continental shelf, biology and chemistry*, ed. R. W. Flint and C. W. Griffin, pp. 9-1–9-62. The 1977 final report to the Bureau of Land Management, Department of the Interior, Washington, D.C. Contract AA550-CT7-11.

Teal, J. M. 1976. Hydrocarbon uptake by deep-sea benthos. In *Sources, effects and sinks of hydrocarbons in the aquatic environment*, pp. 358–371. Washington, D.C.: American Institute of Biological Sciences.

Thayer, P. A., La Rocque, A., and Tunnell, J. W., Jr. 1974. Relict lacustrine sediments on the inner continental shelf, southeast Texas. *Transactions of the Gulf Coast Association of Geological Societies* 24:337–347.

Trefry, J. H., and Presley, B. J. 1976a. Heavy metal transport from the Mississippi River to the Gulf of Mexico. In *Marine pollutant transfer*, ed. H. L. Windom and R. A. Duce, pp. 39–76. Lexington, Massachusetts: D. C. Heath.

——. 1976b. Heavy metals in sediments from San Antonio Bay and the northwest Gulf of Mexico. *Environ. Geol.* 1:283–294.

Trent, L. 1967. Size of brown shrimp and time of emigration from Galveston Bay system, Texas. *Proceedings of the Gulf and Caribbean Fisheries Institute* 19:7–16.

Ward, A. R. 1975. Studies on the subtidal free-living nematodes of Liverpool Bay. II. Influence of sediment composition on the distribution of marine nematodes. *Mar. Biol.* 30:217–225.

Watson, R. L., and Behrens, E. W. 1970. Nearshore surface currents, southeastern Texas Gulf Coast. *Contrib. Mar. Sci.* 15:133–143.

Wieser, W. 1960. Benthic studies in Buzzards Bay. II. The meiofauna. *Limnol. Oceanogr.* 5:121–136.

Wigley, R., and McIntyre, A. D. 1964. Some quantitative comparisons of offshore meiobenthos and macrobenthos south of Martha's Vineyard. *Limnol. Oceanogr.* 9:485–493.

Wohlschlag, D. E., Cole, J. F., Dobbs, M. E., and Vetter, E. F. 1977. Demersal fishes. In *Environmental studies, south Texas outer continental shelf, biology and chemistry*, ed. R. D. Groover, pp. 10-1–10-152. The 1976 final report to the Bureau of Land Management, Department of the Interior, Washington, D.C. Contract AA550-CT6-17.

Wohlschlag, D. E., Yoshiyama, R. M., Dobbs, M., Finley, E., and Vetter, E. F. 1979. Demersal fishes. In *Environmental studies, south Texas outer continental shelf, biology and chemistry*, ed. R. W. Flint and C. W. Griffin, pp. 18-1–18-99. The 1977 final report to the Bureau of Land Management, Department of the Interior, Washington, D.C. Contract AA550-CT7-11.

Wüst, G. 1964. Stratification and circulation in the Antillean-Caribbean basins: Part I. New York: Columbia Univ. Press.

Zobell, C. E., and Feltham, C. B. 1938. Bacteria as food for certain marine invertebrates. *J. Mar. Res.* 1:312–327.

INDEX

Aerobic heterotrophic bacteria. *See* Bacteria
Allochthonous organic materials, 5, 140
Ammonia, 46, 139–141
Amphipods, 100–101, 106
Analysis of variance, 38, 170–172
Aransas Pass Inlet, 17, 23, 26, 28, 54, 109, 147
Aromatic hydrocarbons
 in benthic biota, 124, 127, 129
 lack of, in sediments, 76
 in zooplankton, 61

Bacteria, 88–91, 143–144, 146, 193
 abundance of, 88
 and degradation of hydrocarbons, 88–91
 as a food source, 144–145
 relationship of, with meiofauna, 146
 toxicity to, 90
 in trophic relationships, 145–146
Balanced and unbalanced data, 169–170, 172
Bank stations. *See* Hospital Rock; Southern Bank; Topographic highs
Benthos
 as a community, 139
 as a study element, 83, 144
 See also Demersal fish; Epifauna; Infauna; Meiofauna

Bioaccumulation, 61, 123, 143
Biogenic hydrocarbons, 31, 78, 122
Biomass
 carbon equivalents, 42, 147
 chlorophyll *a* as indicator of, 23
 of demersal fish, 116–117
 determination of, 42–43, 147–150
 of infauna, 150, 152–153
 of microzooplankton, 60, 147
 of neuston, 48, 54–55, 147
 of phytoplankton, 36–43, 137–138, 140, 147
 of phytoplankton as water mass indicator, 23
 of zooplankton, 55–56, 138, 147
Bioturbation, 142, 151
BLM. *See* Bureau of Land Management
Bloom, phytoplankton, 42, 145
Body burdens
 in benthic biota, 122–136, 143
 hydrocarbons, 61–63, 122–130, 143, 188–189
 trace metals, 62–67, 129–136, 189
 in zooplankton, 61–67, 188–189, 143
Brazos complex, 26, 68
Bureau of Land Management, 3, 10

Carbon fixation, 42, 46, 139. *See also* Primary production
Carbon-14 uptake, 38, 41–42
 in nepheloid layer, 46

[232] Index

Catch statistics. See Fishery catch statistics
CFU. See Colony-forming units (CFU), of fungi
Chlorophyll *a*
  as biomass indicator, 23, 36–43
  nannochlorophyll *a*, 36–42, 183, 185
  in nepheloid layer, 34, 46–47
  net chlorophyll *a*, 36–42, 183, 185, 201
  relationship of, with infaunal density, 104, 138
  relationship of, with riverine input, 23–28, 38
  relationship of, with salinity, 23–29
  relationship of, with temperature, 23, 25
  total chlorophyll *a*, 36–42, 184–185, 205
Climate, of study area, 16, 18. See also Meteorology
Cluster analysis
  of demersal fish data, 114–116
  of epifauna data, 107–109
  of infauna data, 97–98
  of neuston data, 55
  of phytoplankton data, 43
  of zooplankton data, 58–59
Coastal zone
  description of Texas', 3
  importance of, 153–155
Colony-forming units (CFU), of fungi, 84–87, 193
Community structure
  of epifauna, 107–113
  of infauna, 96, 98–99, 104–106
Confidence intervals, 160–161, 166–168
Confidence level, 176
Confounded data. See Balanced and unbalanced data
Copepods, in zooplankton, 55–56, 58–59, 185–186, 208
Currents, 28–29, 142
  longshore component, 28
  near shore, 28
  seasonal effects on ecosystem dynamics, 28–29
  wind generated, 28, 54

Data synthesis and integration, 5–6, 12–14, 155
Decapod crustaceans, larval
  in neuston, 48–49
Delta, ancestral, 68–69
Delta ^{13}C
  petroleum pollution and, 73–74
  in sediment, 72–75, 191–192, 212
Demersal fish, 114–122, 195–198
  biomass, 116–117, 195
  diel variation, 116
  diversity, 117
  dominant species, 116–117, 121
  hydrocarbons in, 124–125, 127–130, 196–198
  migration of, 117
  production, 150
  relationship of, to environmental variables, 118–121
  species groupings, 114–116, 118–119
  station groupings, 114–116
  trace metals in, 131–133, 198
Density
  of demersal fish, 116–117, 195
  of epifauna, 109–113, 138, 195
  of infauna, 98–99, 138, 143, 195
Deposit feeders, 96, 145
Descriptive statistics, 163–168
  confidence intervals, 160–161, 166–168
  effect size, 175–176
  harmonic mean, 176–177
  kurtosis, 165
  mean, 163–164
  skewness, 164–165
  standard deviation, 164, 166, 173
  standard error, 173
Detrital food web, 151
Detrital pool. See Detritus
Detritus, 104, 142–143, 145
  based trophic web, 151
Diel variability
  in ammonia, 140–141
  in demersal fish, 116
  in neuston, 48, 55
Discriminant analysis
  of demersal fish data, 118–122
  of infauna data, 101–104
Distributional characteristics of

Index [233]

STOCS ecosystem, 157–177, 178–198
  analysis of spatial and temporal effects, 168–173
  balanced and unbalanced data, 169–170, 172
  comparison with future monitoring results, 173–177
  descriptive statistics, 163–168
  format for presentation of, 157–161
  geographic representation of, 199–200, 200–219
  replicate samples, 163, 168, 177
  sampling error, 167
  selection of variables, 163
Disturbance, 105, 143

Ecosystem
  characterization, 137–139
  definition of, 4
  functioning, 146
Ekman transport of surface water, 22
Environmental disturbance, 97, 143, 153–156
Epifauna, 107–113, 195
  community structure, 107–113
  environmental variables associated with, 107
  faunal provinces, 107
  migration of, 109
  species groupings, 109
  station groupings, 107–109
Equitability
  of epifaunal species' distributions, 109–113
  of infaunal species' distributions, 98–99, 104, 106
  of zooplankton species' distributions, 57–58
Estuaries
  contribution of, to benthic fauna, 109
  influence of, on zooplankton community structure, 58–59
  as part of coastal zone, 3, 16, 23, 137, 139
  as source of input of methane to water column, 31

Ethane, 32–33, 77–79, 179–180, 183
  in nepheloid layer, 32
Ethene, 32–33, 77, 178, 180, 182–183, 207

Faunal provinces, 107
Feeding mechanisms
  of infauna, 106, 145
  of meiofauna, 96, 145
Feeding strategies
  of infauna, 106
  of meiofauna, 145
Fish, in neuston, 49. See also Demersal fish
Fisheries, 4, 83, 97, 109, 139, 142, 146–153, 154
  economic importance of, 154
  pelagic, 149–150
  yields, 150, 152
Fishery catch statistics, 143, 147–148, 150
Food web, 142–146
  detrital, 151
Fungi, 83–88, 193
  community structure, 84
  control of abundance, 87
  degradation of hydrocarbons, 84–87
  inoculum, 87
  population density of, 84
  relationship of, to total organic carbon, 87
  toxicity to, 87

Geochemistry, 4. See also Sediment, chemistry
Geomorphology, 68–69
Gradients. See Spatial variability
Gulf of Mexico
  climate of, 15
  description of, 6–8
  hydrography of, 18–19, 23–29
  trace metal contamination of, 79–82

Harpacticoids, in meiofauna, 94–95, 194, 217
High molecular-weight hydrocarbons (HMWH)
  in benthic biota, 123–129, 196

carbon preference index (CPI), 124, 127
dissolved and particulate, 33, 35, 181
in macronekton, 127–130, 196–198
particulate in water column related to zooplankton, 61
phytane, 123–124, 127, 129, 181
ratios, 123–124
in shrimp, 125–128, 196
SUM HI in sediments, 76–77, 192
SUM LOW in sediments, 76, 192
SUM MID in sediments, 76–77, 192
in water column, 33–35, 181
in zooplankton, 61–62, 188–189
Hospital Rock, 7, 68, 69, 94, 161
Hydrocarbons
in benthic biota, 122–130, 196–197
biogenic origin of, 76, 122
food chain transfer, 123, 125
microbial degradation of, 84, 86, 89–91
in sediments, 76–79, 123, 143, 192
sources in water column, 31
uptake from sediments, 123
in water column, 31–35
in zooplankton, 61–63, 143, 188–189
See also High molecular-weight hydrocarbons; Low molecular-weight hydrocarbons
Hydrography, 18–29, 137
bottom water variability and stability, 21–23, 104–106
circulation, 18, 28–29, 54–55
cross-shelf temperature dynamics, 21
currents, 28–29
isothermal conditions, 21, 139
temperature gradients, 18–19
vertical stratification, 18–19, 22, 87, 140
water chemistry, 29–35
water mass characterization by correlation with salinity and chlorophyll *a*, 23–28
water mass distribution, 18–19, 23–28
Hypersalinity, 16

Ichthyofauna. *See* Demersal fish
Ichthyoplankton, 4, 55, 60, 138, 186
Infauna, 96–107, 145–146, 194–195, 218–219
biomass, 150
community structure of, 98–100
comparison of, to other continental shelves, 106–107
dominant taxa of, 100–101, 106
environmental variables, 101–104
gradational features, 104–106, 138
heterogeneous benthic habitat, 104–106
opportunists, 105
production, 150, 152–153
relationship of, with hydrography, 104–105
relationship of, with primary production, 104–105, 138
relationship of, with sediment texture, 102–106
species composition of, 97, 99–101
species groupings, 99–101
station groupings, 97–101, 102

Kurt. *See* Descriptive statistics, kurtosis

Langmuir cells, 54
Limiting factor
of nutrients, 29
to fungi, 87
Low molecular-weight hydrocarbons (LMWH)
microbial generation, 76–78
in sediments, 76–81
in water column, 31–33, 178, 180, 182
Lutjanus. See Snapper

Macrofauna
epifauna, 107–114
infauna, 96–107
similarities of, to meiofauna, 96, 145

Macroinfauna. *See* Infauna
Macronekton
  hydrocarbon body burdens in, 123–130
  trace metal body burdens in, 129–135
  See also Demersal fish; Epifauna; Shrimp; Snapper
Management, environmental, 153–154
Meiofauna, 91–96, 144–146, 193–194, 214–215
  definition of, 91
  as a food source, 144–145
  nematodes in, 93–95, 194, 216
  permanent, or true, meiofauna, 91–95
  relationship of, to sediment texture, 94
  similarities of, to infauna, 96
  spatial and temporal variability, 92–94, 193, 214–215
  species composition, 94
  temporary meiofauna, 91
  total meiofauna, 94–95, 193
Metabolism, of infauna, 140
Meteorology, 15–18, 137
  air circulation, 17, 87
  air-water interaction, 15–16, 23
  atmospheric pressures, 16–17
  cloud cover, 16
  effect of water temperature on, 15
  hurricanes and tropical storms, 15, 17
  "northers," 15, 17
  rainfall, 16–18, 21
  winds, 16–17, 22
Methane, 31–33, 178, 180, 182–183
  association of, with nepheloid layer, 32, 34
  production of, in water column, 31
  in sediments, 77–78
Microbial generation, 31–32, 76–78
Microbiology, 83–91, 145–146. *See also* Bacteria; Fungi
Microbiota, 145–146
Microtarballs. *See* Tarballs
Microzooplankton, 60, 147
Migration, 109, 113–114, 117, 136

Mississippi River, 6–7, 18, 23, 27–29, 36, 38, 80–81, 107, 137
Models
  conceptual of STOCS, 5–6
  of demersal fish and environmental variables, 118–122
  of infauna and environmental variables, 101–104
  of trophic relations, 143–151
  of water mass dynamics, 27–28
Monitoring
  comparison of, with base line results, 173–177
  effective sample size for, 174–177
  future efforts, 127, 136, 157
Mousse. *See* Tarballs
Mud-water interface. *See* Sediment-water interface
Multiple use, of coastal zone, 154–155

N-alkanes
  in benthic biota, 123–124, 125–128, 196–198
  microbial degradation of, 89
  in sediment, 76–77, 192
  in snapper tissues, 129–130, 196–197
  in water column, 33, 181
  in zooplankton, 61, 188–189
Nannochlorophyll a, 36–42, 183, 185
  carbon-14 uptake and 38, 41–42, 183
National Marine Fisheries Service (NMFS), 4, 147–148, 150
National Outer Continental Shelf Environmental Studies Program, 3
Natural gas seepage, 32, 78, 82
Natural resources, 3, 154–156
Nekton. *See* Macronekton
Nematodes, 93–95, 144–146, 194, 216. *See also* Meiofauna
Nepheloid layer, 32, 34, 46–48, 106, 140–142, 150–151
  carbon-14 uptake in, 46
  chlorophyll levels in, 46–47
  light transmission through, 46–47
  photosynthesis in, 46
  phytoplankton in, 46

Index

and trophic coupling, 151
Net chlorophyll a, 36–42, 183, 185, 201
  carbon-14 uptake, 38, 41–42, 183
Neuston, 48–55
  biomass, 48, 54–55, 147
  definition of, 48
  diel vertical migration of, 48
  euneuston, 55
  facultative neuston, 55
  fish in, 49
  larval decapod crustaceans of, 48–49
  nutrition of, 55
  relationship of, to ichthyoplankton, 55
  species composition of, 48–53
  tarballs in samples, 49, 54–55
95% Empir. CI. *See* Confidence intervals
Nitrate, 29–30, 178, 180, 182
Nitrogen, 29, 139–141. *See also* Ammonia; Nitrate
Nutrient regeneration, 46, 139–142
Nutrients
  in nepheloid layer, 46–48
  recycling of, 142, 145
  reservoir in benthos, 140
  in water column, 28, 29–31, 137, 140, 178, 179, 181–182
  See also Nutrient regeneration

Odd-even preference (OEP)
  OEP HI in sediments, 76
  of particulate hydrocarbons in water column, 61
  in zooplankton, 61, 189
OEP. *See* Odd-even preference
Oil degradation potential
  of benthic bacteria, 89–91
  of benthic fungi, 84–86
Oil spills, 122, 143
Opportunists, 105
Ordination analysis, of infauna, 97–99
Organic matter
  allochthonous, 140
  dissolved and particulate, 33
Oxygen, dissolved, 29–31, 178, 180, 182, 204

PAH. *See* Polynuclear aromatic hydrocarbons
Particulate matter
  association of, with nutrients, 23
  association of, with primary productivity, 23
  association of, with trace metals in zooplankton, 62
  clay particles, 62
  organic, 33
  suspended, 67
  and trace metals, 131
Patchiness
  in benthic fauna, 96
  of phytoplankton, 27
  in sediment texture, 70
  of tarballs, 62
  of zooplankton, 58, 62
Penaeus. *See* Shrimp
Petroleum activities, 4, 48, 54, 76, 84, 97, 154–155
  background levels of hydrocarbons, 122–123
  carbon preference index (CPI) as indicator of, 127
  crude oil importation, 61–62
  oil spills, 122, 143
Phaeophytin, 184, 185, 202
Phosphate, 29–31, 178, 180, 182, 203
Photic zone, 37–40
  half photic zone, 37–40
Physical oceanography, 18–29. *See also* Hydrography
Phytoplankton, 36–46, 183–185
  biomass, 36–43
  carbon-14 uptake by, 38, 41–42, 183
  chlorophyll a as biomass indicator of, 36–43, 138
  community structure, 43
  patchiness, 27
  primary producer biomass, 42–43, 147–150
  relationship of, with zooplankton, 60, 138
  species composition, 43–46
  succession of species, 46
Pollutants, 48, 137, 143, 154–155
  background levels, 122–123
  detection of, 73–74, 127

Index [237]

long-term effects, 122
sources of trace metals, 79–81
See also Oil spills; Petroleum activities
Polychaetes
  as dominant taxa, 99
  in infauna, 99–101, 106, 146
  in meiofauna, 94–96
Polynuclear aromatic hydrocarbons (PAH), in zooplankton, 61
Power. *See* Confidence level
Primary production
  calculation of annual yield, 42–43, 147
  carbon fixation, 42, 147
  carbon-14 uptake, 38, 41–42, 139
  relationship of, with infauna, 104
  relationship of, to particulate hydrocarbons, 33
  and trophic coupling, 142, 147–150
Productivity
  of inner shelf waters, 137–139
  of meiofauna, 96
  relationship of, to dissolved oxygen, 29–31
Propane, 32–33, 77–79, 179, 180, 183
Propene, 32–33, 77, 179, 180, 183, 206
Protozoa. *See* Microzooplankton

Rainfall, 16, 18, 21
Recreation, 154
Recycling, of nutrients, 142, 145
Regression analysis, 59, 170
Replicate samples, treatment of, in data analysis, 163, 168, 177
Respiration, 140
Rhomboplites. *See* Snapper
Rig monitoring study, 10, 127–128
Rio Grande, 6–7, 17, 27–28, 68, 70, 72, 81
Riverine input
  to coastal zone, 16, 17, 21, 81, 136–137, 139
  influence on water mass characteristics, 23–28
  of methane, 31

  relationship of, to chlorophyll *a*, 23–28
  of trace metals, 80–82

Salinity
  of estuaries, 16, 17
  influence of Mississippi River, 18
  influence of riverine input, 18, 21
  relationship of, to chlorophyll *a*, 23–28
  relationship of, to zooplankton, 58–59
  of shelf waters, 18–21
  variability and stability of, 21–22, 105, 107
Sample number, future monitoring, 174–177
Sampling scheme, 7–14, 161–164, 168–173. *See also* Study area; Study design
Sargassum, 48
Seasonal variation. *See* Temporal variability
Secchi depth, 23–24, 62, 178. *See also* Particulate matter
Sediment, 69–72, 189–192
  coarsening, 70–71
  comparison of, to other continental shelves, 96–97
  geographic zones of sediment texture, 72–74
  grain size, 69, 72, 74
  grain size deviation, 72, 74
  inner shelf sands, 69–71, 138
  mid-shelf muds, 70
  outer shelf clays, 69, 71
  and relationship to faunal assemblages, 103–106, 138
  resuspension of, 27, 106, 142
  sorting, 70–71
  texture, 69–72, 103–106, 189–191, 209–210
  transition stations, 72, 74
  transport, 4, 68–69
  variability, 70
Sediment, chemistry, 4, 72–82, 191–192
  delta ^{13}C and total organic carbon, 72–75, 191, 211–212
  hydrocarbons, 76–79, 191

[238] Index

trace metals, 79–82
Sedimentation, 68, 151
Sediment-water interface, 32, 139–140, 151
Seepage
  of methane, 32, 78
  of trace metals, 82
Shrimp, 109, 138, 144–146, 154, 196, 197
  catch statistics, 147, 150
  fishery yield, 83, 143, 147, 150, 152
  hydrocarbons in, 125–128, 196
  as monitors for hydrocarbons, 127, 129, 146
  as part of detrital food web, 151
  as part of fishery conceptual trophic model, 147–152
  survival curve, 150
  trace metals in, 131–136, 143–144, 146, 197
Silicate, 29–30, 178–181, 200
Skew. *See* Descriptive statistics, skewness
SLCO. *See* South Louisiana Crude Oil
Snapper, hydrocarbons in, 127–130, 196–198
Soft-bottom environment, 97
Southern Bank, 7, 68, 69, 94, 161
South Louisiana Crude Oil (SLCO), 84, 86, 90
South Texas Outer Continental Shelf (STOCS), 3, 137–139. *See also* Study area
Spatial variability
  in chlorophyll *a*, 38, 138
  of copepods in zooplankton, 58
  in demersal fish, 114–116
  of demersal fish trace metals, 131, 136, 138
  in distributional characteristics of base line data, 160–163
  in epifauna, 107–113
  of infauna, 97–98, 138
  of meiofauna, 92–93
  of microbial oil degradation potential, 84, 89
  in neuston, 48
  of sediment delta ^{13}C and total

  organic carbon, 72–73
  of sediment low molecular-weight hydrocarbons, 77
  of sediment texture, 69–72
  of tarballs, 54
  in water column hydrocarbons, 31–33
  in water column nutrients, 29–31
  in water salinity, 19–21, 105, 107
  in water temperature, 19–22, 105, 107
  of zooplankton, 55, 138
  of zooplankton trace metals, 62
Species, number of
  in epifauna, 109–113
  in infauna, 98–101
  See also Species richness
Species composition
  of benthic fungi, 87
  of demersal fish, 116–117
  of epifauna, 107–109
  of infauna, 97, 99–101
  of meiofauna, 94–96
  of neuston, 48–53
  of phytoplankton, 43–46
  of zooplankton, 58–59
Species diversity
  of demersal fish, 114, 117
  of epifauna, 109–113
  of infauna, 98–101
  of microzooplankton, 60
  of neuston, 49
  of zooplankton, 57–58
Species richness, in zooplankton, 57–58
STOCS. *See* South Texas Outer Continental Shelf
Study area
  description of, 6–10
  station locations, 7–9, 161–164
Study design, 3–6, 10–14, 161–164, 168–173
  data synthesis and integration, 12–14, 168–173
  principal investigators, 11, 14
  rig monitoring study, 10
  sampling periods, 10, 14, 161
  stations, 7–9, 161–164
  study elements, 11–14
  supplemental studies, 10

Study objectives, 3–6, 155
  characterization of resources, 4–5
  data synthesis and integration, 5–6
Supportive studies
  National Marine Fisheries Service on ichthyoplankton and fisheries, 4
  supplemental studies, 10
  Texas A&M University on topographic highs, 4
  U.S. Geological Survey on sediments, 4
Suspended sediments, 4, 142, 151

Tarballs, 49, 54–55, 61–62, 76
Temperate zone, 106
  phytoplankton activity in, 42
Temperature, air
  coastal, 16–17
  offshore, 17–18
Temperature, water, 15, 18–23
  cross-shelf gradient, 18, 21
  relationship of, with chlorophyll a, 23, 25
  relationship of, to neuston, 55
  surface water, 19
  variability and stability of, 21–22, 105, 107
  vertical stratification, 18–19, 22
Temporal variability
  of benthic bacteria, 88–89
  of benthic fungi, 84
  of carbon-14 uptake, 38, 42
  of chlorophyll a, 23–28, 36–38
  of currents, 28–29
  of demersal fish, 114, 118
  of demersal fish trace metals, 131, 136
  of distributional characteristics of base line results, 160–161
  of epifauna, 110–112
  of infauna, 97–98
  of meiofauna, 93–94
  of neuston, 48
  of phytoplankton, 46
  of salinity, 23–28, 105, 107
  of sediment texture, 70–71
  of shrimp hydrocarbons, 125
  of water column hydrocarbons, 31–33
  of water column nutrients, 29–31
  of water temperature, 21–23, 105, 107
  of zooplankton, 55, 58
  of zooplankton trace metals, 62
Terrestrial origin, of benthic fungi inoculum, 87
Terrigenous input, of particulate hydrocarbons, 33
Tides, 15
Topographic features, of continental shelf, 68–69
Topographic highs, 4, 7, 32, 68–69, 95–96, 97
Total chlorophyll a, 36–42, 184–185, 205
  carbon-14 uptake, 38, 41–42, 183
Total organic carbon
  as a food source, 144
  relationship of, to benthic fungi, 87–88
  in sediments, 72–75, 191, 211
Tourism, 154
Trace metals
  anthropogenic sources, 80–81
  atmospheric sources, 136
  in benthic biota, 129–136, 198
  clay particle adhesion, 62
  in demersal fish, 131–133, 198
  in sediments, 79–82, 131, 143–144
  toxicity, 129
  in zooplankton, 62, 64–67, 143–144, 189
Transfer efficiency, 147, 150
Transition stations
  for infaunal species' composition, 97–103
  for sediment texture, 72–74
Transmissometry, 32, 34
Trophic coupling, 83, 92, 142–153
  carbon transfer, 147, 149–150
  nickel as food web tracer, 143–144, 146
t-test, 173
Turnover ratio, 147, 151

Ubiquitous fauna
  demersal fish, 116
  fish taxa in neuston, 49
  infauna, 99

Unbalanced data, 169–170
Upwelling, of deep Gulf waters, 22, 38
U.S. Geological Survey, 4

Variables, analyzed for temporal or spatial differences, 163, 168–173. *See also* Distributional characteristics of STOCS ecosystem

Water, chemistry, 29–35. *See also* Hydrography
Water column. *See* Hydrography
Wind, 16, 17, 22
  influence of, on currents, 28

Zooplankton, 55–67, 185–189
  biomass, 55–56, 138, 147, 186
  community structure, 57–58
  copepods, 55–56, 58–59, 185–186, 208
  density of, 55–56
  effects of salinity on, 58–59, 60
  fecal pellets of, 61, 143
  hydrocarbon body burdens in, 61–63, 188–189
  microzooplankton, 60
  patchiness of, 58
  production, 58, 147
  relationship of, with ichthyoplankton, 60
  relationship of, with phytoplankton, 60, 138
  spatial and temporal variability, 55, 58
  trace metal body burdens in, 62–67, 189